CONTENTS

Dedication

For Delia and Michael

Acknowledgements

In writing *Sex, Botany and Empire*, I relied heavily on books and articles by other historians. I am particularly indebted to the work of Richard Drayton, John Gascoigne, Lisbet Koerner and Londa Schiebinger. In addition, I should like to thank Harriet Guest and Anne Secord for their helpful suggestions, and also Sujit Sivasundaram, who commented on a draft of the whole book.

About the author

After gaining a degree in physics at Oxford University, Patricia Fara ran a company producing audio-visual material on computing before switching to the history of science. A specialist in 18th century England, she now teaches at Cambridge University and is a Fellow of Clare College. A regular contributor to radio and TV programmes such as *In Our Time*, she has published widely on the history of science. Her prize-winning *Science: A Four Thousand Year History* was translated into nine languages, and other highly acclaimed books cover diverse topics including Isaac Newton, women in science, electricity, Erasmus Darwin and the First World War.

· Chapter 1 ·

The Three Ss

That curiosity which leads a voyager to such remote parts of the globe as Mr B— has visited, will stimulate him when at home ... As nature has been his constant study, it cannot be supposed that the most engaging part of it, the fair sex, have escaped his notice; and if we may be suffered to conclude from his amorous descriptions, the females of most countries that he has visited, have undergone every critical inspection by him.
Town and Country Magazine, September 1773

Harriet Blosset was rich, beautiful, and delighted to be watching an opera with her fiancé, a wealthy young Lincolnshire landowner called Joseph Banks. She probably never forgot the date – 15 August 1768. A messenger arrived at the theatre to inform Banks that he should report immediately to Plymouth, where James Cook was waiting for him on the *Endeavour* to set sail for the Pacific Ocean. During dinner at her family home that evening, Banks drank heavily while she vowed tearfully to live quietly in the countryside. Left behind to embroider waistcoats for her absent lover, Miss Blosset plunged into depression and became obsessed with death. In contrast, Banks rapidly recovered from their separation; he was away for nearly three years and had a marvellous time.

The *Endeavour* expedition to Tahiti and Australia is central to this book because it changed not only Banks's life but also the pattern of British science. The ingenuous young botanist became President of London's Royal Society, where he reigned over an international scientific empire for more than 40 years. Although he was never allowed to forget the sexual slanders about his early exotic adventures in the Pacific region, Joseph Banks became a powerful administrator who convinced the British government that investing in scientific research would benefit the country's commercial and imperial expansion. More than any other single individual, Banks welded together the Three Ss – Sex, Science and the State.

* * *

Backed by the Admiralty and the Royal Society, Cook sailed the *Endeavour* towards Tahiti in order to observe a rare astronomical event, the Transit of Venus, when the planet crosses in front of the Sun. Cook's ship was packed with shiny brass instruments under the care of navigators who had been specially trained to use them. This was the first of Cook's three round-the-world voyages, and one of the earliest scientific expeditions to be funded by the state. Banks, however, was paying for himself and his seven servants and assistants, because collecting foreign plants was not the sort of project that attracted government money. Banks was no scholar – he had struggled through the classics syllabus at Harrow and Eton and failed to complete his degree course at Oxford. But he had been fascinated by botany since he was a child, and he pulled strings at the Admiralty so that he too could travel on the *Endeavour*. Banks was enthusiastic about the new system for classifying plants and animals that had been introduced by the Swedish expert Carl Linnaeus, now an elderly man but still much admired. As Banks set sail for the Pacific, he probably reflected on Linnaeus's youthful expeditions to the Arctic regions, and perhaps fantasised that one day he would supplant Europe's great botanic emperor – a dream that did, in fact, come true.

How appropriate, many people must have thought, that Venus, the planet of love, was to be observed from the island of Tahiti. Discovered only the previous year by a British expedition led by Samuel Wallis, Tahiti already represented an exotic paradise for Europeans, an earthly Garden of Eden. According to travellers' reports, the fine climate and fertile soil nurtured an uncorrupted natural society, a people who lived in harmony, totally free of the decadent vices plaguing Western civilisation. Above all, the Tahitians supposedly suffered none of the sexual inhibitions that so restricted English enjoyment; on the contrary, the gratification of erotic desire was seen as one of life's major objectives.

When Wallis sailed close to the island in his *Dolphin*, he was immediately struck by the stunning scenery, but later marvelled even more at the social customs. At first, the crew members and the local inhabitants were mutually suspicious. For several days, they negotiated language barriers while also deciding how much they could trust one another. Not surprisingly, the Europeans resorted to their guns, and several Tahitians died before an effective system of bartering was discovered, in which sex was the main unit of currency. As canoe-loads of lovely young women were paddled around the *Dolphin*, even the chronically ill members of the sick bay rallied and begged permission to go ashore.

The Tahitian women initially allowed themselves to be enticed with trinkets, but as time went by they cleverly raised the price of seduction. There was no iron on Tahiti, so a metal that was commonplace in England there became a valuable commodity as the women insisted on receiving longer and longer nails. The ship's master disdainfully recorded events as though he himself were not participating: 'all the Liberty men carried on a trade with the Young Girls, who hade now rose their price for some days past, from a twenty or thirty penny nail, to a forty penny, and some was so Extravagant as to demand a Seven or nine Inch Spick.' Although the carpenter swore that he was guarding his supplies closely, he probably made a handsome profit, since the *Dolphin* began to disintegrate as nails and cleats were pincered out of the wood.

A few months later, a French expedition landed on the island, and their ecstatic reports reinforced Tahiti's reputation as an idyllic Utopia of free love, an Elysian paradise currently available on Earth to ordinary mortals. By the time that Cook and Banks arrived in 1769, the islanders were suffering from sexually transmitted infections which they called 'Apa no Britannia' – the British disease. Cook struggled to maintain discipline among his men, ordering them back to the ship when they disappeared on shore. But he also had an anthropological zest for joining in local practices, and he ate roast dog, shaved himself with a shark's tooth, and stripped to the waist so that he could attend traditional ceremonies. Suppressing his own emotions, he dispassionately noted that erotic dancers 'keep time to a great nicety' (although he did lose count as one naked soloist twirled around); Cook also survived a strange Sunday which opened with a Christian service and closed with what he drily called 'an odd scene' of public sexual intercourse.[1]

By the standards of his age, Cook was a tolerant man who tried hard to observe and analyse without judging. Inevitably, his measured neutrality sometimes wavered. Tones of moral outrage crept into his journals, and he found it impossible not to condemn the promiscuity he witnessed. Banks had no such qualms – he was there to enjoy himself. Although Cook might remain a spectator, Banks revelled in being a participant, confessing his escapades with the same self-congratulatory candour as James Boswell (who naturally longed to visit Tahiti as soon as he heard that it rivalled London for romantic delights). The female dancers found it advantageous to keep their eccentric visitors happy, and Banks readily believed that they singled him out for special attention. After their leader had pirouetted naked in front of him and presented him with some cloth, he 'took her by the hand and led her to the tents accompanied by another women her freind'. Sometimes he joined in the dancing himself, wearing only a loin cloth but not 'ashamd of my nakedness for neither of the women were a bit more coverd than myself'.

His special 'flame', he wrote, was Otheothea, the personal attendant of a high-ranking woman called Purea – or Queen Oberea as she was mistakenly called by the Europeans, who misheard her name and elevated her rank because they were insensitive to fine social distinctions between people they

6

lumped together as an inferior race. During the *Dolphin*'s visit, Purea had taken over Wallis's social agenda. She distracted him from perpetrating further carnage among the islanders by entertaining, massaging and feeding him; convinced of her devotion, he rewarded her with lavish presents. By the time that the *Endeavour* arrived, Purea had been defeated in a civil war, but she still tried to manipulate the island's uninvited guests, even organising a ritual copulation ceremony for Banks to observe.

Banks's subsequent notoriety as a Pacific rake hinged on an incident in Purea's canoe, where she had invited him to spend the night. Because of the heat, he discreetly explained, 'I strippd myself' and yielded to Purea's suggestion that she look after his clothes. When he woke up a bit later, he discovered that his elegant white jacket and waistcoat with silver frogging had disappeared, along with his pistols and gunpowder. After handing over his musket to one of Purea's men to keep guard, Banks settled back down to sleep, but was forced to emerge the next morning with a borrowed robe wrapped round his shoulders, 'so that I made a motley appearance, my dress being half English and half Indian'. Purea, Banks concluded sadly, had probably colluded in the theft of his garments and weapons. After trying unsuccessfully to negotiate some pigs in compensation, he shamefacedly trailed back to the *Endeavour*.[2]

* * *

Banks's exploits in the Pacific were a gift for gossip columnists and satirists. For years after he arrived back in England in 1771, caricatures, pamphlets and articles mocked his sexual activities during the voyage. The Purea episode featured in many of the vicious poems written in what now seem painfully contrived rhyming couplets. This is a typical example:

> *She sinks at once into the lover's arms,*
> *Nor deems it vice to prostitute her charms;*
> *'I'll do,' cries she, 'What Queen's have done before';*
> *And sinks, from principle, a common whore.*

In addition to being savaged for exploiting the Tahitian ruler, Banks was criticised for refusing to honour his engagement to Miss Blosset (although her family did successfully extract a substantial financial settlement from him). He was accused of opting for science rather than sex. Harriet Blosset's ex-paramour had, critics sniped, been seduced by 'the elegant women of Otaheite [Tahiti] ... but *she found her lover now preferred a flower, or even a butterfly to her superior charms'*. Satirical poets voiced similar complaints about his abandonment of Purea. In one long defamatory epistle, Oberea (Purea) tries to lure back her British botanist by converting herself into a luxuriant plant and weaving her 'wanton foliage round thy hand'. He has, she laments, deserted her for science – 'at least ... spare one thought from Botany for me', she begs.[3]

Banks became known as the 'Botanic Macaroni' (Figure 1). The term 'Macaroni' was originally coined to denigrate the aristocratic youths who had acquired continental manners during their Grand Tour to Italy, but it became a more general term of abuse for deriding foppish young gentlemen who adopted ridiculous extremes of stylish clothing. The label was laden with sexual contempt. A Macaroni, sneered one journal, is 'neither male nor female, [but] a thing of the neuter gender ... It talks without meaning, it smiles without pleasantry, it eats without appetite, it rides without exercise, it wenches without passion.'[4] Just as plants were grouped into families, and people into tribes, so too the caricaturists identified different types of Macaroni, classifying them by the streets they paraded in, or the occupations they devised to fritter away their time.

Dressed in an ultra-fashionable coat and wig, the Botanic Macaroni carries a sword, by then no longer the essential prop of an elegant gentleman but the sarcastic (and highly symbolic) attribute of a Macaroni too effeminate to know how to use one. His right leg is swathed in bandages, a unique early

Figure 1. 'The Botanic Macaroni' (1772), by Matthew Darly. (© The British Museum.)

reference to the gout that would later make him an invalid. Ineffectually smiling and clutching his magnifying glass, Banks is a botanical libertine whose excessive desire for women has been replaced by an obsessive preoccupation with plants.

Botany may now seem a harmless scientific pursuit, but in the late 18th century it was fraught with sexual allusions. When satirists jeered at Banks for offering an exceptionally large plant to Queen Oberea they were not being particularly original, even though in his case the joke carried extra bite because Banks really was a botanist. Throughout the Enlightenment period, lewd poems graphically compared women's bodies with geographical features such as hills, rivers and creeks, while plants provided pornographic analogies for the sexual organs of both men and women. Legacies of this erotic botanic intensity survive in words such as 'defloration', the vibrant flower paintings of the American artist Georgia O'Keeffe, and the floral place settings in Judy Chicago's feminist art installation, 'The Dinner Party'.

To make matters worse, even the scientific language of botany was saturated with sexual references. Banks ardently supported the controversial Linnaean system of classification, which relied on counting the numbers of male and female reproductive organs inside flowers. To describe different groups of plants, Linnaeus had used extraordinary terms like 'bridal chamber' and 'nuptials'. For prudish Britons, this sexualised version of nature verged on the pornographic, and battles over botanical textbooks resembled current debates about allowing children to watch violent videos. Self-appointed moral guardians of society declared that they wanted to protect young women from the corrupting influence of botanical education. They clamped down on mixed flower-gathering expeditions, and sanitised floral vocabulary by introducing meaningless euphemisms. By allying himself with Linnaeus's supporters, Banks opened himself up to widespread insinuations about his sexual activities.

* * *

One of the most successful parodies of Banks and his sexual prowess was *Mimosa: or, The Sensitive Plant*, which appeared with Banks's name on the title page in 1779, seven years after his return from Tahiti. The book was published anonymously, but several Enlightenment writers appreciated the metaphoric potential of this plant that visibly shrinks and grows. It is hard to imagine modern adults laughing at, let alone buying, a long poem that slanders the sexual proclivities of prominent aristocrats through botanical innuendo. Nevertheless, such satires are rewarding to study because humour provides a marvellous entrée into other cultures. The opening sentences of the preface convey *Mimosa*'s flavour: 'The world will determine with what justice I dedicate the SENSITIVE PLANT, to a Gentleman so deeply skilled in the science of Botany … The plains of *Otaheité* … rear that *plant* to an amazing height … and Queen Oberea, as well as her enamoured subjects, feel the most sensible delight in *handling*, *exercising*, and *proving* its virtues.'

Presumably the poem's readers did not find such puns tedious, even when repeated several times in different versions. Four pages later, the anonymous author approached the conclusion of his dedication to Banks. 'Men of science, with equal *ardour*, have entered on the same task, and you, Sir, stand foremost in the list of those, who, anxious for the propagation of the PLANT, have explored worlds unknown before, and brought home to your native land, discoveries of its virtues, and relations of its vigour.'[5] Hardly subtle – yet significantly, the writer was drawing on familiar clichés of geographical and botanical pornography to colour scientific exploration with imperial overtones of possession, domination and exploitation.

This *Mimosa* dedication neatly ties together the Three Ss – Sex, Science and the State. Its visual counterpart is Figure 2, another Macaroni caricature of Banks. As in 'The Botanic Macaroni', the

Figure 2. 'The Fly Catching Macaroni' (1772), by Matthew Darly. (© The British Museum.)

redundant sword and elaborate feathers hint at sexual ambiguity. Banks had scoffed at the young gentlemen who wanted to complete their education in Europe; his Grand Tour, he had declared, would be one round the whole world. Although he took every opportunity to enjoy himself, Banks regarded himself not as a tourist but as a traveller. Like Linnaeus before him, he wanted to capture the world by classifying it scientifically. And here Banks is being mocked for these imperial pretensions, his feet uncertainly straddling the two halves of the globe. In order to enlarge his scientific collection, he vainly strives to catch a butterfly, symbol of triviality. The caption sneers:

I rove from Pole to Pole, you ask me why,
I tell you Truth, to catch a ___Fly!

Like modern political cartoons, these Macaroni caricatures are superficially funny but also hint at deeper criticisms of social structures. Cook's voyage of exploration was no naïve search for scientific truth. The astronomical and botanical observers on board did, of course, make many new discoveries, but the voyage's backers had provided funding to meet commercial and political objectives. Banks complained about the dishonesty of royal hosts who stole the clothes of their sleeping guests, but apparently had no compunctions about theft on an international scale. While he was taking over the indigenous women and plants, Cook was securing Pacific territories for the British nation.

There is no single correct way of interpreting the past: as people try to make sense of their own lives, they repeatedly create new versions, new memories. Or, as the Danish philosopher Søren Kierkegaard put it, life is lived forward but understood backward. Scientists like to browse through earlier centuries and pick out glorious ancestors whose illustrious achievements seem to presage their own success. To boost their own position, they construct stories that celebrate science's inevitable progress, as if a torch of truth were handed on from one great man to the next (or, very occasionally, a woman). In these triumphant tales, Carl Linnaeus appears as a botanic forefather who introduced a major system of classification that is still in use today; Joseph Banks, on the other hand, features merely as an adventurous explorer, an assiduous disciple who used Linnaeus's schemes to catalogue the plants and animals he collected.

This heroic style of telling history may be traditional, but it leaves many questions unanswered. To start with, it does not explain how, why and when science and its applications became so fundamental in society. For centuries, the top subjects were theology and the classics, and the balance only slowly started to tip towards science and mathematics. Historians often neglect the 18th century because it lacks famous figureheads such as Isaac Newton or Charles Darwin, yet this was a crucial period when science started to become established and gain prestige. Along with their Enlightenment contemporaries, Linnaeus and Banks fought hard to establish that scientific knowledge was valid and valuable.

Men like Newton and Darwin are commemorated as great heroes who – supposedly – produced revolutionary theories, yet soared above the petty demands of everyday affairs. In simplistic visions of the past, science develops in a make-believe world inhabited by disinterested scientists with only one objective – to uncover truth. Reality, of course, is different. Scientific investigators are driven not only by their genuine fascination with nature, but also by other motives – power, money, fame. Linnaeus in Sweden and Banks in Britain illustrate how scientific research is intertwined with commercial development and imperial exploitation.

Science's history is too often converted into an exciting race between intrepid investigators who are competing to reach the peaks of truth. Such adventure stories may be enthralling to read, but they are not much help in explaining how science has become integrated within our daily lives. Modern science depends on industrial and government financing, and it is naïve to divorce the growth of scientific knowledge from the development of its importance. When science's social significance is incorporated within its history, the linked stories of Linnaeus and Banks unfold very differently. Viewed in retrospect, Linnaeus clung to an older vision of imperial domination that ultimately failed as an economic and scientific experiment. In contrast, Banks emerges not as a disciple, but as the prophet of a scientific empire that came to rule the world. Without enthusiastic preachers like him, the theories of our traditional scientific heroes – Newton's gravity, Darwin's evolution – would not have become common knowledge, indispensable components of our scientific and technological world.

Banks's innovations placed science at the heart of Britain's trading and political empire. Linnaeus may be the scientific star of botany, but Banks had a greater long-term impact. As an autocratic administrator, he perhaps lacks the glamour of Newton, Darwin or even Linnaeus, yet he too deserves to be commemorated as one of science's founding fathers. President of the Royal Society for over 40 years, Joseph Banks ensured that science and the British Empire flourished and expanded together. He forged an interdependent relationship between science and the state that endures today.

· Chapter 2 ·

The Scientific Swede

Observe too what irregularity passes
From the want of distinction of Sexes & Classes ...
Can Marriage made public & Marriage
  clandestine
The same common bed with strict decency rest in
Shall a Couple as constant as Darby and Joan
In a basket with libertine flaunters be thrown ...
No no my gay empire will sooner dissever
And my Colonists claim independence for ever
    Hand-written verses inside Joseph Banks's copy
      of A System of Vegetables by Carl Linnaeus
        (translated by Erasmus Darwin), 1783

'God created and Linnaeus organised' – so quipped a colleague of Carl Linnaeus (1707–78), the son of a Swedish pastor who introduced a new way of classifying plants and animals. Linnaeus made no life-changing discovery, and was often mocked as an ill-educated provincial boor. Nevertheless, he rapidly became celebrated as one of science's great heroes because he invented a revolutionary method for labelling plants that was easy to use. His new

'Language of Flowers' was, he boasted, so straight-forward that even women could understand it. For the first time, botanical enthusiasts from all back-grounds could learn a simple way of identifying flowers – and his classification system is still in widespread use today.

The 18th century is often dubbed 'The Age of Classification', and Linnaeus was the classifier *par excellence*. By 1799, over 50 different systems were available, but Linnaeus's was the one that survived. In his *Geography of Nature*, he divided living organisms into different groups and sub-sets arranged in an orderly five-tier pattern of categories – classes, species and so forth. From now on, he said, every plant and animal should carry its own unique two-part label. Lemon trees, for instance, were called *Citrus limon* to distinguish them from their close relatives, orange trees, or *Citrus aurantium*. And Linnaeus also coined a new term to describe human beings – *Homo sapiens*, or wise man.

Because Linnaeus's system has been in use for over 200 years, it often seems that this way of classifying plants and animals must be the natural or right way to do it. But modern scientists are still arguing about its merits, and his scheme was enormously controversial when he first proposed it in 1732. Many of his rivals were trying to work out God's original blueprint for the universe, and they accused Linnaeus of choosing an arbitrary plan rather than one that was divinely ordained. He was also criticised for building an elaborate structure on the basis of relatively unimportant features. Earlier botanists had tried to group plants by characteristics such as the colour of their flowers or the shape of their leaves, but Linnaeus decided to order plants numerically according to their reproductive organs. Surprising though it might seem, it had been nearly the end of the 17th century before naturalists realised that plants reproduce sexually. Even though many plants are hermaphrodites, which carry both male and female parts, Linnaeus settled on this sexual dichotomy for organising the plant world.

As his model for this supposedly objective system, Linnaeus turned to human relationships. The prejudices of Enlightenment Christian moralists are built right into the heart of this scientific plan for plants, which Linnaeus outlined by using romantic words such as 'bride' and 'marriage'. In his anthropomorphic scheme, the most basic division is between male and female – exactly the same distinction as in the highly chauvinistic society of late 18th-century Europe. Linnaeus gave priority to male characteristics; in other words, he imposed the sexual discrimination that prevailed in the human world onto the plant kingdom. His first level of ordering depends on the number of male stamens, while only the sub-groups are determined by the number of female pistils.

From the dominant position enjoyed by Linnaeus and his male contemporaries, this way of dividing the plant kingdom carried a huge advantage: it made his arbitrary organisation of plants appear as though it were natural, even God-given. Linnaeus had mapped human society onto the botanical world, but from then on men of science could argue in reverse. Since sexual hierarchies prevail in nature, male supremacy must also – so the distorted logic runs – be appropriate for people; this argument conveniently forgets how this sexual ordering was inferred from society in the first place. Through this closed loop, Linnaean classification not only mirrored social prejudice, but also reinforced it.

Figure 3 shows how Linnaeus organised plants into 24 classes according to the number of male stamens in the flower. By counting the female pistils, he then sub-divided each of these classes into less important orders. Linnaeus was a profoundly religious man who believed in the sanctity of marriage, yet his text reads like a parody of a Mills and Boon novel: 'The flowers' leaves ... serve as bridal beds which the Creator has so gloriously arranged, adorned with such noble bed curtains, and perfumed with so many soft scents that the bridegroom with his bride might there celebrate their nuptials with so much the greater solemnity. When now the bed is so prepared, it is time for the

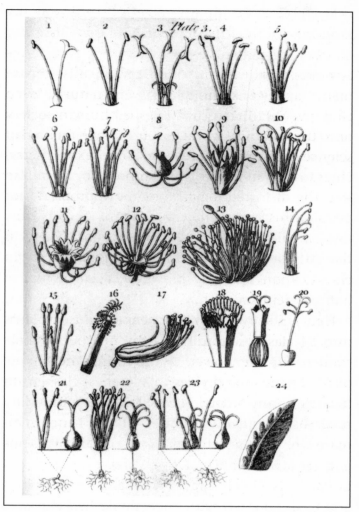

Figure 3. Linnaean classes, based on a drawing by Georg Ehret illustrating Carl Linnaeus's *Systema natura* (1737). From James Lee, *Introduction to Botany* (1760). (By permission of the Syndics of Cambridge University Library.)

23

bridegroom to embrace his beloved bride and offer her his gifts.'[6] Critics were quick to denounce this sexual vocabulary.

Paradoxically, the man who introduced eroticism into botany was a home-loving pastor who refused to let his daughters learn French in case they lost their appetite for housekeeping. He equated sexuality with marriage rather than promiscuity, and regarded women as wives and caregivers rather than as individuals with their own desires and ambitions. Linnaeus called plants in the first class *monandria*, from the Greek for 'one man'. He nicknamed his own wife a monandrian lily – a virgin with a single husband. However, many orders of plants had unequal numbers of stamens and pistils, and so could not possibly correspond to conventional marriages. Linnaeus described these unorthodox arrangements with words like 'concubine' and 'clandestine marriage'. Perhaps he thought of himself as belonging to the third order of the first class, a single man with three wives – his monandrian lily, his Parisian botanic illustrator, and his deepest love, Dame Nature herself.

* * *

Linnaeus spent most of his adult life at the University of Uppsala. This small town is now only about an hour by train from Stockholm, but was then considered a provincial backwater. Largely self-taught, Linnaeus himself became a local curiosity rather like the Sami servants (then called Lapps), African slaves and exotic animals that belonged to the royal collections he looked after. Born into a family of country curates, Linnaeus cultivated a deliberately scruffy appearance and conversed only in southern Swedish or schoolboy Latin with a Nordic accent. He was, however, an excellent self-publicist.

Figure 4 is a good example of how Linnaeus manipulated his public image. Originally designed to impress a rich patron, this portrait shows Linnaeus in traditional Sami clothes, as though he were an intrepid voyager freshly returned from the hostile Arctic regions. He had, in fact, assembled this costume for touring through Europe to back up his colourful, exaggerated travel tales. Linnaeus commissioned several versions of this picture, but it is very deceptive.

Just as Banks and his English companions looked down on the Tahitians, so metropolitan Swedes regarded the Sami people as an inferior race. By adopting their clothes for this portrait, Linnaeus was masquerading as an exotic indigenous person, a tactic that had the effect of reinforcing his true status as an imperial possessor. Any Sami could (if foolhardy enough) have told Linnaeus how ridiculous he looked. His beret, a present from a Swedish

Figure 4. 'Carl Linnaeus returned from Lapland. Thirty years old 1737'. Engraving by H. Meyer from the painting by Martin Hoffman (1737). (National Library of Medicine/Science Photo Library.)

tax collector, was suitable for women in the summer. His winter fur jacket, which he had bought in Uppsala, came from a different region, and his reindeer leather boots were made not to wear but to export for rich, gullible southerners. His shaman's drum – another gift – was an illegal possession. To complete the look, Linnaeus dangled assorted tourist souvenirs from his belt.

The white flower on his chest, which he had named after himself, became a Linnaean trademark because he made sure that it was always included in his portraits. He was so attached to this *Linnaea borealis* – Linnaea of the north – that he brewed it to make Lapp tea (disgusting, reported his son). This small Arctic plant advertised Linnaeus's 1732 voyage to Lapland, which in reality was only a brief and timid foray towards the north. Then 25 years old and funded by the Uppsala Science Society, he set out in a braided wig and elegant leather trousers, well provided with two night-gowns as well as his microscope, goose quills and plant pressing equipment. Also secreted away in his luggage were some useful maps and travel diaries compiled by previous explorers: he kept very quiet about those. Although Linnaeus was away for several months, he spent only eighteen days in Lapland and never even crossed the Arctic Circle into the polar north. Since he was paid per mile, for his final report to the Society he more than doubled the distance he had travelled, drawing maps with long but fictional detours.

Despite these dubious tactics of self-promotion,

Linnaeus was sincere in his aspirations. He wanted to make his country self-sufficient by cultivating foreign plants at home. Sweden had lost much of its empire and – unlike Britain – had little prospect of acquiring further territories overseas. Europeans, argued Linnaeus, were spending money on importing goods from Asia, yet had little to sell in return (apart from guns): far better, he claimed, to cease being forced to rely on farflung lands. Linnaeus's success in growing Europe's first banana plant helped to win support, and he convinced the government to invest in his projects for taming colonial crops so that they would grow in Scandinavia. Linnaeus dreamed of transforming the national economy. Rice paddies in Finland, cinnamon groves in Lapland, tea plantations in the Baltic: in Linnaeus's futuristic visions, Sweden would enjoy the same luxuries that Britain and Holland obtained from their foreign empires.

After his expedition to Lapland and a three-year stay in Holland, Linnaeus came back to Sweden in 1738 (when he was 31), got married, and never left the country again. Trained as a doctor, he initially specialised in treating syphilis, but three years later he was appointed professor of medicine at Uppsala University. For the next 40 years, he consolidated his academic position, publicised his ideas about classification, and tried to reform the Swedish economy through his agricultural recommendations. His lectures attracted students from all over Europe, including England.

God was central to Linnaeus's plans. A strict Lutheran, he studied the Bible closely. On his interpretation, human beings had a double divine mission – to look after the world, and to exploit it for their own benefit. By deciphering God's laws of nature, Linnaeus taught, naturalists could take advantage of the world's riches. They had a duty to investigate plants not just out of scientific curiosity, but also in order to find ways of turning them into medicines, food or shelter. Every country had been blessed with useful plants: science's task was to discover and cultivate them. Of course, in Linnaeus's view, white European men were best suited to achieve this goal. Through understanding how God supervised the universe, they could learn how to manage the Earth and its inhabitants. For him (and many others) imperial rule was a responsibility imposed by God.

Linnaeus regarded himself as a second Adam. In the Garden of Eden, Adam had named the animals placed there by God. Working at Uppsala, Linnaeus re-designed the University's botanical garden to make it an image of God's creation, a miniature paradise on Earth (Figure 5). Charles Darwin had not yet been born, and like many of his contemporaries, Linnaeus believed not in change and evolution, but in stability. In their understanding

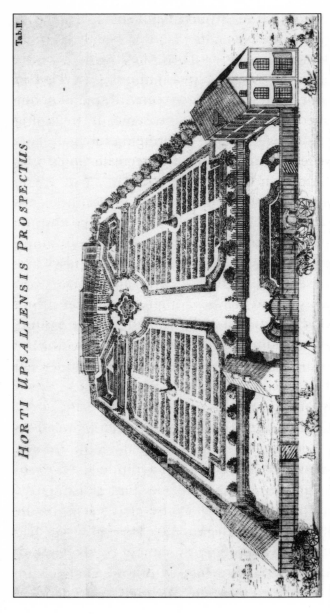

Figure 5. 'Hortus Upsaliensis (Linnaeus's garden at Uppsala). From Carl Linnaeus, Hortus Upsaliensis (1745). (By permission of the Syndics of Cambridge University Library.)

of the Bible, all the world's plant species had originally been present in the Garden of Eden, which Linnaeus envisaged as a small island at the Equator. Subsequently, he explained, although species had diversified to suit different environments, they remained fundamentally the same. By reversing this scattering process, and bringing foreign plants to Sweden, Linnaeus aimed to recreate God's original Garden in Uppsala.

Continuing the tradition of centuries, Linnaeus gave his garden clear boundaries to separate its internal order from the post-Fall wilderness beyond. The neat beds, divided into annuals and perennials, were arranged strictly in accordance with Linnaeus's system, as though he were planting out God's own classification scheme for visitors to observe, admire and learn from. He divided the grounds into quarters because four was a special number. The four rivers of the Garden of Eden corresponded to the four great rivers of the world, and Linnaeus's method of grouping human beings was based on the Four Continents: Europe, Asia, Africa and – the most recent addition – America.

As curator, Linnaeus lived just outside this earthly paradise, in the house at the front right-hand corner. Like the garden, his home was also designed as a miniature museum of God's creation. Birds nested in the branches stacked up against the walls, which were papered with botanic prints and hung with portraits and dried plants. Shells dangled from the ceiling, while monkeys and racoons scampered between geological specimens, scientific instruments and stuffed animals. Among the ornaments jostling for attention were his fictitious Lapp costume, and china decorated with his own heraldic flower, *Linnaea borealis*.

Displaying the same reforming zeal as his preacher father, Linnaeus set out to win converts to his system. His handbooks resembled Lutheran almanacs, divided into twelve chapters and 365 aphorisms – each day had its own botanical text. Determined to make his ideas easily accessible, Linnaeus wrote clearly, ensured that his books were cheap, and gave practical instructions about collecting, labelling and growing plants. As he continually refined his system, Linnaeus published prolifically, writing in Swedish for local people and in simple Latin to reach international audiences. With his impressive reputation, he attracted visitors from all over Europe.

For years, Linnaeus led groups of up to 300 people on hikes through the Uppsala countryside, persuading them how easily they could now identify plants. Linnaeus classified people with the same enthusiasm as flowers, organising his followers with military discipline into groups ruled by a hierarchy of commanders under himself as general. Unpunctuality was punished, and an informal uniform was mandatory. After a successful outing, Linnaeus would march at the head of his botanical troops as they paraded into town, brandishing their trophies and accompanied by a band of musicians. Eventually, the Rector of the University intervened, accusing students of neglecting their duties. In unintentional self-parody, he banned these enjoyable excursions on the grounds that 'we Swedes are a serious and slow-witted people; we cannot, like others, unite the pleasurable and fun with the serious and useful'. Linnaeus was devastated, partly because these tours were extremely lucrative: making botany available to the masses was a profitable business.[7]

Banks always intended to pay homage to Linnaeus. Before he knew that he would be sailing on the *Endeavour* with Cook, he told a friend not to be annoyed if Banks 'Sacraficed every Consideration to an opportunity of Paying a visit to our Master Linnaeus & Profiting by his Lectures before he dies who is now so old that he cannot Long Last' (from the vantage point of a 24-year-old, 60 does seem ancient).[8] But later, after Banks had consolidated his own status, he broke his promise. The older man must have realised that Banks was gradually ousting him from his position as Europe's most powerful botanic emperor.

* * *

Ensconced in his chaotic house at the edge of his Uppsala garden, Linnaeus established a scientific empire whose tentacles stretched around the globe. While he retained control of the central hub, he sent out his best students – his disciples, he called them – on botanical pilgrimages. They had two complementary missions. Fully trained in his methods, they were instructed to search for unusual plants and bring them back to be acclimatised and so help make Sweden self-sufficient; conversely, they were to spread the Linnaean gospel among the international botanical community. In the long run, they proved far less successful at persuading plants than botanists. Linnaeus's transplanted crops mostly withered away, whereas his classification system thrived internationally because of its simplicity.

Linnaeus's briefs for his apostles resembled trading manifestos rather than botanical descriptions. He wanted to collect plants not for their interest as rarities, but for their value as commercial commodities. Although he aimed to reform the Swedish economy, Linnaeus's view of economics was different from the one in use today. For Linnaeus, economics was not about global finance, but about tapping in to God's own natural economy on a local basis. No mathematician, he knew nothing of modern concerns to balance supply and demand or boost industrial production.

Instead, he tried to husband the resources of the natural world, which he believed God had designed as one great circulating economy – a common older use of the word. Linnaeus envisaged the world as a self-regulating hierarchy in which natural economies functioned at several different levels. Each animal or plant formed its own little balanced system. Together, these made up the larger economy of the local environment, which in its own turn made up part of a nation's economy. Linnaeus wanted to restore the world to the paradise that God had originally created, by enabling each country to produce all the goods it needed for its own economic survival.

Some economists argued that God had scattered His riches around the Earth in order to encourage international trade, but Linnaeus was convinced that God intended Sweden to prosper by providing all its needs within its own borders. He sent off his disciples to scour foreign countries for useful materials, even recommending them to smuggle goods into Sweden if necessary. From China, he ordered natural products such as a tea bush, a sample of soil for making porcelain, rubber, palm fruits and (perhaps a tactful afterthought) some live goldfish for the queen; the postscript smacked of industrial espionage – how, the Academy of Sciences wanted to know, did the Chinese smelt zinc? Similarly, a traveller to America was told to bring back some especially tasty rice, a selection of mulberry trees so that Sweden could cultivate silk worms, and oxen with long hair for spinning and weaving.

Linnaeus professed to be undeterred by the most obvious obstacle to his ambitious plans – the Swedish weather. The trick lay, he claimed, in fooling the plants by gradually getting them accustomed to colder and colder climates: after starting them off in southern Sweden, they could be moved northwards a bit at a time. Any casualties – and there were many – could be explained away by blaming the weakness of the individual specimen; the species as a whole, Linnaeus argued, could be tamed through this process of slow adaptation.

Linnaeus also developed a convenient theory of Alpine climates. According to him, plants that grow on the top of mountains, near the tree line, could be successfully moved to the desolate areas in the Arctic north. The indigenous nomads, he decreed, could be converted into agricultural labourers to tend the tea plantations, saffron meadows and cedar forests that would flourish among the glaciers and lichen-covered boulders. Although some rebellious students did object that his proposals were scientifically invalid, most of the 186 dissertations that Linnaeus supervised corroborated his findings – hardly surprising, since he wrote many of them himself.

In retrospect, the outcome of Linnaeus's projects is obvious – Sweden is still not renowned for its production of silk, coffee, rice and other foreign crops. Linnaeus's schemes, however solidly buttressed by his scientific theories, were not successful. Many of the exotic samples from abroad never even reached Sweden – lacking effective protection, lizards, peacocks and tender plants failed to survive the long, stormy sea voyage to the north. Although Linnaeus did manage to make tropical plants flourish briefly in his garden at Uppsala, they rarely lasted for long. The mortality rate was also high among his disciples, many of whom died of tropical diseases or drowned at sea.

Sadly, Linnaeus missed out on some marvellous opportunities. Potatoes, which are hardy and packed with calories, might have been his big success story, but Linnaeus thought they were poisonous. After all, he argued, they are related to deadly nightshade, and even pigs dislike them. Although there were no Swedish recipes for cooking potatoes, in 1748 one woman had a brainwave – they could be used for making wig powder and aquavit. As a reward for this inspiration, she became the only woman to be elected to the Swedish Academy of Sciences before the 20th century.

Scientists often claim to provide an objective picture of nature, but Linnaeus's science was clearly driven by his political agenda. Because he wanted to make Sweden into a nation 'that doesn't have to rely on foreigners', he devised – and sincerely believed in – botanical theories that justified his bid to make Sweden self-sufficient.[9] Already famous as a classifier, Linnaeus convinced the government and the Swedish scientific societies to invest in his ideas because his theories about the economy of nature corresponded to national ambitions for a self-sufficient financial economy. Persuaded by Linnaeus, Sweden embarked on a series of experiments with a double aim – to try out a new type of economic system, and to alter the existing distribution of useful plants. But these experiments failed on both counts, and Linnaeus's reputation declined. By the early 19th century, his garden at Uppsala was overgrown with weeds and the greenhouse stood in ruins.

* * *

In Britain, upright botanists were appalled by the sexual implications of Linnaeus's classification system. Taking advantage of the opportunity to make a feeble pun (another example of how fashions in humour have changed), they sneered at his 'florid' style. Linnaeus had clearly spelled out the analogies between the reproductive organs of flowers and people. 'The calyx is the bedchamber', he explained in 1735, 'the filaments the spermatic vessels, the anthers the testes, the pollen the sperm, the stigma the vulva, the style the vagina'. Such explicit explanations seemed scandalous – 'too smutty for British ears', one critic spluttered – but especially so for one half of the population. As one clergyman protested, 'Linnaean botany is enough to shock female modesty'.[10]

The new Swedish system was slow to catch on. Although botanists were familiar with it by the middle of the 18th century, it was still controversial 50 years later, despite the efforts of Linnaeus's campaigning disciples. Even Linnaeus's friend Philip Miller, the botanist in charge of the Chelsea Physic Garden, took many years to be converted. When the Physic Garden was eventually rearranged to follow a Linnaean scheme, the gardeners grumbled about the extra work. Many of the new plants, they protested, 'are of foreign Production, of tender natures and especially such as are raised from seeds, in Hot Bedds; and require frequent Shifting and changing of situation'.[11]

One of the first institutions to adopt the Linnaean scheme was the British Museum, which opened in 1759 after Sir Hans Sloane, President of London's Royal Society, donated his famous collection of almost 80,000 objects to the nation. Like other early collecting enthusiasts, Sloane had gathered together under one roof a huge variety of items which now seem to have little in common.

Under the influence of the Royal Society, they were regrouped in a way that feels far more familiar.

In older displays of curiosities, wonders of nature – such as abnormal foetuses or exceptionally large shells – jostled for attention with human artefacts like valuable coins and ancient musical instruments. At the British Museum, these natural and artificial curiosities were physically separated into the two halves of the new building – one side for books, manuscripts and medals, the other for natural history specimens. For many years, stuffed giraffes loomed over the top of the staircase, and it was only in 1881 that the animals and plants moved to their present site in South Kensington. Sloane had always opposed Linnaeus, but the first director of the new Museum insisted on adopting Linnaean classification for the gardens as well as for the plants and animals displayed inside. He even recommended that cabinets for storing dried plants should be constructed with 24 drawers – one for each of Linnaeus's classes – following the specifications that Linnaeus himself had laid down.

Gradually, English translations and commentaries on Linnaeus's ideas started to appear. In the second half of the 18th century, Linnaean botany became very fashionable, so that botanists made money by conducting field trips while popular magazines encouraged people to take up plant collecting as a hobby. One of the most influential books was by James Lee, owner of a large nursery in London for cultivating foreign plants (it was on a visit to admire Lee's rare plants that Banks had first met 'the fairest amongst the flowers' – Harriet Blosset, later to be discarded).[12] Lee corresponded with Linnaeus, who named a plant after him – *Leea*. By drawing on his practical experience, Lee produced a guide that made it easy to apply Linnaean ideas to English flowers.

Women were often targeted as purchasers of these new publications, and female enthusiasts included Dorothy Wordsworth (the poet's sister) as well as Queen Charlotte, who took refuge from George III's bouts of insanity by retreating with her daughters to Windsor Park, where she 'sits in a very small green room which she is very fond of, reads, writes, and botanizes'. Writers catering for this genteel audience were trapped in a dilemma. On the one hand, studying flowers seemed an ideal pastime for women – not too taxing mentally, a gentle occupation that could be carried out peacefully at home but also involved some therapeutic exercise. On the other hand, botanic vocabulary vibrated with sexual innuendo.

William Withering, a physician famous for curing heart problems with medicines made from foxglove (*Digitalis*), opted for the sanitised version of Linnaean botany. In his best-selling textbooks, he translated contentious words into harmless but meaningless English equivalents such as 'chives' and 'pointals'. Explicitly writing for women, he aimed to make botany 'as healthful as it is innocent' so that it 'leads to pleasing reflections on the beauty, wisdom, and the power of the great CREATOR'.[13] With Withering's bowdlerised botany, from which sex and Latin had been expurgated, women could discuss flowers safely without being accused either of sexual impropriety or of pedantry.

Other botanists disagreed with this approach. After all, they commented, concealing the basic rationale of Linnaeus's system took away much of its point. One of Withering's most outspoken opponents was his friend Erasmus Darwin, also a doctor, whose ideas on evolution were later picked up by his far more famous grandson, Charles. Drawing on the great dictionary compiler Samuel Johnson for advice, Darwin faithfully translated Linnaeus by retaining many of his Latin terms and making clear the sexual basis of his method. Astutely, he dedicated his version to the new President of the Royal Society, Joseph Banks. Flattered, Banks lent Darwin books and checked his work.

Still more controversially, in 1789 Darwin eventually plucked up the courage to publish his long poem celebrating Linnaean sexuality, *The Loves of the Plants*. Romantic, erotic, and enhanced by learned footnotes, it was an immediate success. Darwin's *Loves of the Plants* reinforced the close connections between botany and sexual promiscuity. Borrowing episodes from mythology, in over 1,700 lines of verse Darwin converted Linnaeus's system into rhyming couplets and unabashedly revelled in its erotic implications. In his pornographic paradise, gods and goddesses cavort freely in every imaginable combination. All sorts of female stereotypes – the virtuous virgin, the timorous beauty, the laughing belle, the dangerous siren – reflect the desires and prejudices of Georgian gentlemen. This is how Darwin described the plant *Collinsonia*, which has two male stamens and one female pistil:

> *Two brother swains, of COLLIN's gentle name,*
> *The same their features, and their forms the same,*
> *With rival love for fair COLLINIA sigh,*
> *Knit the dark brow, and roll the unsteady eye.*
> *With sweet concern the pitying beauty mourns,*
> *And sooths with smiles the jealous pair by turns.*[14]

The *Encyclopaedia Britannica* found this second class of plants particularly obnoxious. 'A man would not naturally expect to meet with disgusting strokes of obscenity in a system of botany', it expostulated, 'But ... obscenity is the very basis of the Linnaean system'.[15] The Reverend Richard Polwhele agreed. He retaliated with *The Unsex'd Females*, a long poetic parody of Darwin which attacked sexualized botany and also vilified liberated women. Like many of his conservative contemporaries, Polwhele wanted women to be passive, docile and domestic, and he reserved his bitterest vitriol for Mary Wollstonecraft, who was campaigning for women's education. *Collinsonia* provided his ideal image for satirising Wollstonecraft. Sniping at 'botanic bliss', Polwhele salaciously portrayed her as 'the Fair-one' simultaneously tending two lovers – a dig that struck home as rumours circulated about Wollstonecraft's love affairs:

> *– But hark! lascivious murmurs melt around;*
> *And pleasure trembles in each dying sound ...*
> *Thrill'd with fine ardors Collinsonias glow,*
> *And, bending, breathe their loose desires below ...*
> *Bath'd in new bliss, the Fair-one greets the bower,*
> *And ravishes a flame from every flower.*[16]

Botany was sexy, dangerous – and big business. When the publishing entrepreneur Robert Thornton produced a new edition of Lee's *Introduction to Botany*, he boosted sales by writing a gossipy introduction about Banks and Blosset. In a project that lasted for years, Thornton hoped to make his fortune by organising an extensive series of lavish botanical illustrations to accompany a Linnaean text. Commissioning famous artists to produce elaborate coloured plates of exotic flower arrange-

44

ments, Thornton devised several strategies for marketing his wares, including selling part works and a set of fine engravings called *The Temple of Flora*. As his fortunes plummeted, in desperation he ran 'A Royal Botanical Lottery', in which the first prize included Linnaeus's portrait (Figure 4). So one self-promoter was marketing another – but Thornton's ambitious schemes proved a financial failure.

More modest ventures did succeed in popularising the new system. By the beginning of the 19th century, a wide range of publications was making Linnaean classification available not only to learned gentlemen, but also to less-educated people, including women and workers. Cleaned-up versions of Linnaean classification meant that botany became one of the very few sciences recommended for girls to study, and mothers were encouraged to take their daughters on healthy rambles to gather flowers. Female authors started writing simplified primers, so that young children – boys as well as girls – were familiar with the basic principles of classification.

As Linnaeus had boasted, his scheme was not necessarily restricted to the middle classes. Around Manchester, groups of weavers set up informal Botanical Societies which met in the local pub (although members were fined for turning up drunk). After piling up the plants they had collected on the table, these artisan botanists used Linnaean textbooks to identify their specimens; through rote repetition, even illiterate labourers could learn the names. Eminent scientists searching for rarities came to rely on these local experts who carried out the hard work of collecting unusual specimens in the countryside. One weaver was so determined to master Linnaeus's 24 classes that he wrote them out 'on a sheet of paper and fixed it to my loom-post, so that when seated at my work, I could always have opportunities of looking it over'.[17]

· Chapter 3 ·

The British Botanist

*The Political State of a Nation may be compard
to a Tree, the Roots of which are the Farmers,
the lower Branches the Retale Traders, the upper
ones the Manufacturers, the Flowers & Fruit to
the Gentry & Nobility; if we cease to supply the
Roots with Manure, the Branches Leaves Flowers
& fruit must fade and wither, but in fact the more
effectually the Root is nourishd the more vigorously
the whole that is above it, will thrive & prosper.*
Letter from Joseph Banks to the Prime
Minister, 10 February 1815

Not everybody liked Joseph Banks. James Boswell,
Samuel Johnson's biographer, thought that he
resembled 'an elephant, quite placid and gentle,
allowing you to get upon his back or play with his
proboscis', but some of the Fellows of the Royal
Society had a very different view of their leader.
'The President is incurably sick with the lust of
domination', his opponents railed in a pamphlet
intended to topple Banks from power; 'he imagines
himself born to rule (Good God! how little do men

know themselves!) and cannot perceive that he has neither the intellectual nor the moral qualities of a ruler.'[18]

Just as opinions differed while he was alive, so too Banks's reputation fluctuated after his death. The early obituaries lavishly praised his contributions to science, declaring that 'To the nation he has bequeathed ... a name that it will never cease to cherish while science is encouraged or respected'. This was an unfortunate forecast. Since then, science has flourished, but very few British people have heard of Banks – unless, that is, they have visited Australia, where Banks became a national hero.[19]

Banks (1743–1820) was a man of multiple identities – aristocratic landowner, botanic explorer, scientific administrator, to name just three. His father died when he was a toddler, so Banks and his younger sister Sarah both knew that when he was 21 he would inherit their country estates, along with a more than comfortable income of £6,000 a year. As some comparisons: Philip Miller, the botanist in charge of the Chelsea Physic Garden, earned £50 a year, the first director of the British Museum £200; and for commanding the *Endeavour*, James Cook was rewarded with an exceptionally high wage of five shillings a day (about £90 a year).

An idyllic childhood of huntin', shootin' and fishin': Banks was expert at all three. In contrast, at school at Eton he was an unenthusiastic pupil, constantly in trouble with his teachers until – according to mythical versions of his life – he experienced a Pauline conversion at the age of fourteen, and fell in love with botany. He paid local medical women to teach him their expertise, and purloined a herbal text book he found in his mother's dressing room. For the first time ever, Banks was seen reading during his free hours.

In the 18th century, there were only two English universities – Oxford and Cambridge – and the entry qualifications were money and religion, not brains. Banks went to Oxford, where many students devoted far more energy to drinking, riding and gambling than to their academic work. But for the first few years, Banks's uncle reined in his expenditure, and although he never did take his degree, Banks studied botany and developed the plant collection he had started at Eton. After he came into his fortune, he commuted between his rooms in Oxford, the Lincolnshire estate which he managed throughout his life, and his central London flat close to the British Museum. Here he met Daniel Solander, who had been one of Linnaeus's favourite disciples before he decided to settle in England and break off his ties with Sweden. Banks also resumed his earlier visits to admire Sloane's collection and the Chelsea Physic Garden, where Linnaeus's friend Miller had gathered together 5,000 plants from all over the world.

When he was 23, Banks embarked on his first international expedition, a nine-month voyage to Canada on a man-of-war with another wealthy landowner's son, Constantine Phipps, who later became an aristocratic Arctic explorer. Battling with sea-sickness, Banks enthusiastically trawled for seaweed and jellyfish, and kept a detailed journal which reveals his literary skills as well as his combination of scientific and sporting interests: 'In the Evening went out Fishing had no sport at all at the harbours mouth tho there seemd to be abundance of Small Trout saw no signs of Large ones Killd today a Kind of Mouse ... which Differs scarce at all From the English Sort.'

As when he later sailed with Cook, Banks was a passenger on a political mission, a scientific hanger-on who had little control over where he went. Interspersed with his exclamations of delight at finding crabs and honeysuckle are his comments on the disputes between French and English settlers and the local inhabitants. In spite of problems with fever, storms and foreign food, Banks returned with several hundred specimens (including a porcupine who 'after sulking for three or four Days ... begins to Eat & I have great hopes of Carrying him home alive').[20]

In recognition of his achievements, Banks was soon invited to join the Royal Society. Like many of the other Fellows, he was a wealthy *bon viveur* who took his scientific interests very seriously. What better way, he thought, to use his inherited wealth than on a scientific expedition with the *Endeavour*? Pouring money into this private project, Banks assembled equipment and companions – including Solander and another Linnaean apostle – and the following year set off for the Pacific Ocean.

* * *

Banks's departure was a turning-point in Harriet Blosset's life, but at the time few people realised how important the *Endeavour* voyage would be for British science and imperial expansion. It was only after he got back that the significance of his adventure started to emerge. Welcoming the glory, Banks contributed to his own publicity campaign.

Within a few months, Banks's uncle commissioned a splendid portrait by Benjamin West, reproduced here as Figure 6. To please fond relatives, young gentlemen on their Grand Tour to Italy often posed for pictures showing them standing in a heroic posture among classical ruins. Although Banks's Grand Tour souvenir was painted in a studio, it showed an international traveller. A curtain of rich red bark material is looped back from the artificial backdrop, and Banks is surrounded by artefacts he had brought from the Pacific. Perhaps referring obliquely to the time he had lost

Figure 6. 'Joseph Banks' (originally entitled 'A Whole Length of a Gentleman with a New Zealand mantle round him'). 1773 mezzotint by John Raphael Smith after the 1771–2 portrait by Benjamin West. (© The British Museum.)

52

his clothes in Oberea's canoe, Banks is wrapped in a fine Maori cloak which half conceals his blue tailored outfit with its gold buttons and frilly white cuffs.

Like the portrait of Linnaeus (Figure 4), this is an image of imperial possession: it resembles the contemporary portraits of soldiers who adorned themselves in the costumes of the native Americans they had conquered. These cultural cross-dressers were boasting about surviving their foreign experiences without being overpowered by them. The rare objects that Banks has purloined, such as the unusual Tahitian adze near the bottom right, are displayed as evidence of the exotic people he has encountered. Later they would find their way into museums for British people to marvel at.

Even at this early stage in his career, Banks knew about the power of pictorial propaganda. This portrait was displayed at the Royal Academy, and was then engraved for sale so that people could hang it on their walls or place it in their print folders for visitors to admire. Banks presents himself as a young imperial adventurer with a keen eye for trading opportunities. By pointing towards his gold-coloured cloak, which is made of flax and fringed with the finest dog hair, he emphasises that New Zealand is a good source of the flax that the British Navy so desperately needed for making sails. The national benefits of scientific research are further advertised by the book of botanical drawings by his left foot, which lies open at a picture of flax.

By demonstrating the commercial rewards of international expeditions, Banks advertised the value of a new social category: the scientific explorer. He helped to transform the stereotype of the English male traveller from the foppish aristocrat degenerating on his Grand Tour to the masculine hero risking his life for the sake of England and of science. However, Banks himself never ventured outside Europe again. His plans to accompany Cook on a second voyage fell through, apparently because the Navy disagreed with Bank's plans to adapt the ship to suit his own requirements. 'Mr Banks', they complained, 'seems throughout to consider the Ships as fitted out wholly for his use ... and himself as the Director and Conductor of the whole; for which he is not qualified and if granted to him would have been the greatest Disgrace that could be put on His Majesty's Naval Officers.'[21]

Although he stayed at home, Banks subsequently became responsible for ensuring that many other young men embarked on scientific expeditions financed by the state. By persuading the government to fund international voyages, Banks ensured that this new heroic role model was perpetuated. In Mary Shelley's *Frankenstein*, Captain Walton epitomises the scientific explorer. On his way to the

North Pole, Walton writes a letter to his sister reminding her of their childhood library, packed with exciting tales of earlier expeditions like those of Cook and Banks. In what seems almost a parody, Walton boasts how he sacrificed himself to the cause of science: 'I voluntarily endured cold, famine, thirst, and want of sleep; I ... devoted my nights to the study of mathematics, the theory of medicine ... My life might have been passed in ease and luxury; I preferred glory to every enticement that wealth placed in my path.'[22]

Banks played a vital role in creating Walton's glorious ideals, even though he failed to live up to them himself. After the three-year voyage that established his reputation as an explorer, Banks settled in to the comforts of his London clubs and his Lincolnshire estate – even his size reflected his enthusiasm for metropolitan pleasures. Apart from a brief trip to Iceland, he scarcely journeyed abroad again (although he did once visit France to help an alcoholic friend escape his creditors). But his publicity agents worked hard, presenting Banks in the press as an intrepid traveller who had rejected 'the allurements of dissipation to explore scenes unknown, and to cultivate the most manly qualities of the human heart'.[23]

* * *

Like many 18th-century gentlemen, Banks believed

in enjoying himself. The philosopher David Hume remembered staying at a country inn with Banks, Phipps (by then a Lord) and the Earl of Sandwich, the First Lord of the Admiralty. Accompanied by 'two or three Ladies of Pleasure', Hume wrote, they 'had pass'd five or six Days there, and intended to pass all this Week and the next in the same Place; that their chief object was to enjoy the trouting Season'. This was in 1776, when the British colonies in America were declaring their independence. Hume found it rather strange that the head of the Admiralty should be spending three weeks fishing while the empire dissolved.[24]

Within a few years of returning from the Pacific, Banks had become the confidant of King George III, ensconced himself in a comfortable London house, and married a woman even richer than himself. No legitimate children appeared, but after his sister Sarah moved in, Banks luxuriated in a harmonious *ménage à trois* (supplemented by lovers outside the matrimonial home). For a couple of months every summer, the three members of the Banks family – together with their numerous servants – decamped to Lincolnshire, where Banks concentrated on managing his country estates. His household was renowned for its hospitality rather than its elegance. Banks was a blunt, forthright man, while Sarah courted gossip by dressing eccentrically. Although she shared his passion for collecting, she lacked formal education and was precluded from attending all-male gatherings. Like other clever women, she could experience science only vicariously through her brother.

When the President of the Royal Society resigned in 1778, Banks managed to rally enough support to be elected, even though he was only 35 years old. Over the next 42 years, by exerting an authoritarian grip over the Society, Banks made science central to British culture. His Soho Square house became the hub of an international scientific empire. Here Banks gathered together his countless natural history specimens, which later formed the basis of the British Museum's collection, and invited colleagues like Solander to his famous breakfasts, where scientific men could discuss their latest acquisitions. He also corresponded with men all over the world: the 20,000 letters that survive (out of an estimated 100,000) are striking evidence of how hard Banks worked. Like Linnaeus, he evidently enjoyed cataloguing, since he devised a meticulous filing system that enabled him to retrieve papers efficiently whether he was in his London or his Lincolnshire home.

Secluded in university departments, modern academics are divided into followers of the Arts and the Sciences. In contrast, Banks and his contemporaries were immensely sociable men who mixed work and pleasure across the disciplines. Experts in

Greek and mathematics were recruited to bolster the intellectual status of the Royal Academy of Art, while admirals, aristocrats and artists mingled at the Royal Society, which was described as 'the first literary society in the world'.[25] As well as Samuel Johnson, James Boswell and other eminent authors, Banks's close friends included the artist Joshua Reynolds, President of the Royal Academy. Between them, the two Presidents, Banks and Reynolds, dominated metropolitan élite life. They visited each other's houses, attended meetings at the Royal Society as well as exhibitions at the Royal Academy, and belonged to the same clubs.

Much of their time was spent in exclusively male company. Unlike France, where women ran influential *salons* attended by both sexes, in England mixed groups for intellectual discussion were rare. Although Banks had several mistresses, he was said to be inept at conversation in female company. Reynolds painted a pair of collective portraits showing the Dilettante Society, an exclusive men's dining club that sponsored artistic trips abroad. These two pictures illustrate how Banks and his aristocratic male friends combined their appreciation of fine antiques with gentlemanly connoisseurship of women and wine. As they raise their glasses, they gaze at two objects held up for admiration – a beautiful gem and a lady's garter.

The Royal Society was split into two major

factions – Banks's supporters and his enemies. Although Banks was expert at manoeuvring rich and influential patrons onto important committees, he published virtually no academic papers. He won the vote for President because he was a wealthy naturalist with aristocratic connections, but many members would have preferred a more intellectual man who was interested in physics, mathematics and the technological applications of science. Banks's critics repeatedly accused him of being an ignorant dilettante who loaded the Fellowship with his own friends and undemocratically imposed administrative alterations to strengthen his rule.

Throughout his long reign, Banks was on guard against the dissensions that threatened to split the Society apart. Rather than appeasing his opponents with minor concessions, he preferred to consolidate his own position and maintain the Society's traditional activities. Afraid of the links between scientific and political revolution, he adopted a conservative approach which came to seem distinctly old-fashioned after 40 years on the President's throne. Although many Fellows clamoured for change, Banks boasted to an elderly colleague that 'my Freinds of the Royal Society have not been infected with the Mania of Reform'.[26]

Despite his eminence, Banks was often satirised and was never allowed to forget the sexual slurs against him. Figure 7 shows a caricature by James

The great South Sea Caterpillar, transform'd into a Bath Butterfly.

Description of the New Bath Butterfly, taken from the Philosophical Transactions for 1795: This Insect first crawld into notice from among the Weeds & Mud on the Banks of the South Sea; & being afterwards placed in a Warm Situation, by the Royal Society, was changed by the heat of the Sun into its present form — it is notic'd & Valued Solely on account of the beautiful Red, which encircles its Body, & the Shining Spot on its Breast; a Distinction which never fails to render Caterpillars valuable.

Figure 7. 'The great South Sea Caterpillar, transform'd into a Bath Butterfly' (1795) by James Gillray. (© The British Museum.)

60

Gillray, which was published in 1795 to mock Banks's recent award from the King of the Red Ribbon of the Order of the Bath. This honour was normally reserved for diplomats and soldiers, and Banks proudly displayed his star and red sash in all his subsequent portraits. Personal vanity, no doubt, but this distinction also underlined how he had made science important for the state. For official ceremonies, Banks wore his costume of pink and white silk with a splendid ostrich feather hat, trappings that implied the endurance of a fine English tradition (deceptively so, since this Order dated from 1725).

Modernisers sneered at Banks for his conservative attitudes, but dyed-in-the-wool reactionaries felt that he represented a new scientific methodology threatening to change Britain for ever. The shell near Banks's left shoulder is a *bonnet rouge* of the French revolutionaries, and his clothes are predominantly red, white and blue. In the brightly coloured original, pink and yellow shells contrast strongly with his blue and green wings. Recalling the earlier pornographic satires, Gillray emphasised the corrupt sexuality and phallic qualities of this Banksian caterpillar rising from the Tahitian soil to be transformed into a Bath butterfly. In his long hand-written caption, Gillray parodied the learned language of the Royal Society's journal by describing how this 'New Bath Butterfly ... first crawl'd into notice from among the Weeds & Mud on the Banks of the South Sea ... it is notic'd and Valued Solely on account of the beautiful Red which encircles its Body, & the Shining Spot on its Breast; a Distinction which never fails to render Caterpillars valuable.'[27]

By now Banks was an influential establishment figure, and the crown at the source of the sun's rays alludes to his close association with George III. As a colleague remarked facetiously, Banks had acquired the position of 'His Majesty's Ministre des affaires philosophiques [Minister of scientific business]'.[28] Running the Royal Society, scrutinising the Mint, overseeing the Board of Agriculture and the Royal Greenwich Observatory – Banks forged tight bonds between science and the state. Sending convicts to Australia, organising expeditions to Africa, cultivating foreign crops in Kew Gardens – Banks was an active committee man who played a vital role in the development of Britain's expanding empire.

* * *

But savage caricatures and poems continued to circulate. At one stage, 'The Fly Catching Macaroni' (Figure 2) had been on display in a print-seller's shop window in the Strand, and other critics took up this theme of Banks foolishly flitting after rare butterflies. The most famous attack was by the pseudonymous Peter Pindar, who made Banks the central victim in one of his long gossipy poems parodying public figures. To illustrate Pindar's verses, the caricaturist Thomas Rowlandson supplied a frontispiece mocking Banks's botanic breakfasts in the 'Flea Room' in Soho Square. Instead of politely conversing about their specimens, the eminent (named) guests are greedily guzzling exotic animals:

Most manfully their masticators using,
Most pleasantly their greasy mouths amusing,
With coffee, buttered toast, and bird's nest chat.[29]

Banks worried about his public image. It seems likely that he tried to suppress a particularly vicious lampoon, *The Philosophical Puppet Show*, since extremely few copies survive. This savage verse satire jeered at Sir Joseph Margin (terrible pun!) and his sycophantic supporters. Although published anonymously, it was obviously written by one of the mathematicians who mutinied against Banks's autocratic rule in 1784 and threatened to split the Royal Society apart. At one level this dissension arose from differences between physicists and natural historians, between modern quantification and old-fashioned virtuosity. But it was also about class. One of Banks's spies reported that his opponents saw the rancorous debates as 'a struggle of the men of science against the Maccaronis of the Society'. The renegades were, he noted, gaining support mainly among 'the inferior members' who felt 'that they ought no longer to be rode by your Maccaroni gentlemen'.[30] Banks managed to quell this rebellion by wheeling in his allies, but dissatisfaction simmered throughout his long reign.

In addition to squashing disapproval, Banks also took positive steps to boost his reputation, monitoring how he was portrayed and boasting about the sales of flattering engravings. He tried to be discreet about this media control. In response to an enthusiastic proposal for a commemorative Sèvres vase, he disingenuously insisted: 'I do not feel as if Vanity was a Prominent trèe in my character.' Nevertheless, he vituperatively rejected this 'intended Brittle Compliment' because he disapproved of its illustrations.[31]

One of Banks's favourite artists was Thomas Phillips, a member of the Royal Academy and now most famous for his theatrical portrait of Lord Byron in an Albanian costume. Like West's portrait of Banks (Figure 6) or the earlier one of Linnaeus (Figure 4), this is an example of cultural cross-dressing. Byron's own poet's shirt peeps out from beneath his gold and red velvet coat, but for authenticity his spectacular turban should have been replaced by a small red cap. Byron posed carefully to consolidate his reputation as Britain's most successful (and notorious) Romantic poet. Similarly, Banks made sure that Phillips's portraits advertised his own authority.

Phillips painted several different versions of Banks, all of them based on his first one showing Banks dressed as the nation's most eminent scientific adviser – the President of the Royal Society (Figure 8; unfortunately this engraving, although of high quality, cuts off some elements of the original painting). As well as twice copying his own original, Phillips altered the accessories to present Banks in other guises, including a country landowner and the President of the Horticultural Society. These portraits went on public display. However, no prints were ever made of privately-commissioned pictures which revealed that Banks was an elderly invalid who wore special boots and was confined to his chair by gout.

Phillips's portrait (Figure 8) epitomises the second major role model that Banks helped to establish – the scientific administrator. Leaning forward from his Presidential throne, Banks exudes authority and seems to be listening attentively before pronouncing judgement. He has chosen to wear a luxurious black jacket, set off by lace at his neck and wrists, and a white silk waistcoat decorated – appropriately for a botanist – with small coloured flowers. This quasi-regality is emphasised by the coat-of-arms above his head, the ceremonial

Figure 8. 'Joseph Banks'. 1812 engraving by Niccolo Schiavonetti after the 1808 portrait by Thomas Phillips. (© The British Museum.)

mace and inkstand lying across the front of the picture, and the overwhelming redness of the picture – the diagonal slash of Banks's Order of the Bath ribbon is matched by the leather chair-back behind him and the elaborate velvet cushion on which he rests his right hand.

When Banks learnt that this picture was going to be exhibited at the Royal Academy, he wrote to Phillips and specified some important details: self-presentation mattered. Tactfully, he asked for the pamphlet in the bottom centre to be by the picture's Spanish commissioner, a mathematical astronomer who wished to thank Banks for his support. However, he insisted that the far more prominent paper in his hand should be a recent lecture by Humphry Davy, the electrical and chemical experimenter who strongly opposed Banks's autocratic rule and would eventually succeed him as President. Banks knew that his mini-empire at the Royal Society was in danger of splitting up, but he obstinately resisted the calls for change by younger Fellows. One of his opponents described the Society as 'a volcano augmenting its power', and Banks was determined to hold down the potential explosion.[32] By displaying himself in this portrait as Davy's patron, Banks symbolically asserted his dominating rule over the Society's dissident factions.

Phillips was keen to make money by having the portrait engraved for sale. Banks's long letter to Phillips is a masterpiece of diplomatic, deceptive self-abnegation. He was evidently delighted with the suggestion, but worried that his critics would accuse him of vanity. So at first, he modestly demurred: 'A man like me who has never medled in Politics, & who Cannot, of Course, possess a Squadron of Enthusiastic Friends, is not likely to Sell a dear Print', he politely if unconvincingly protested. However, in the closing paragraph, he revealed his true interests by recommending a good engraver.[33]

Banks's friend Reynolds had made portrait-painting respectable, but many artists still regarded this national speciality with contempt. Portraiture, sneered a cynic, is 'one of the staple manufactures of the empire. Wherever the British settle, wherever they colonise, they carry and will ever carry trial by jury, horse-racing, and portrait-painting.'[34] He could well have included Banks in this list of imperial products. Of the three copies that Phillips made of Figure 8, two – including the original – are now in Australia, where they are displayed in major libraries, in Sydney and Canberra, which are open to the public; in contrast, the version still in England hangs in a private committee room at the Royal Society.

Despite his enormous importance for science and empire, Banks is virtually unknown in Britain, whereas in Australia he has been converted into a national hero. In contrast with this antipodean adulation, Banks's 42 years as President gained him very little recognition in Britain. Keen to appear progressive, the Royal Society tried to brush up its image in the 19th century by appearing more democratic. When Davy and his colleagues finally took over, they wanted to minimise the importance of a man they dismissed as an old-fashioned autocrat who knew nothing about modern physics and mathematics.

Now that national heritage has become big business, British people seem more interested in claiming Banks back from Australia. In 1986, West's portrait (Figure 6) suddenly resurfaced after having mysteriously disappeared for 120 years. In the auction room, tension mounted as two competing Australian bidders forced the final price up to a stunning £1,815,000. However, Britain's National Portrait Gallery objected to the purchaser's success, claiming that the picture should not be exported because Banks was British and it was Banks who had 'realised the potential of the newly discovered lands and set out to promote the concept of settlement and colonisation'. This imperial argument proved so convincing that funds were raised to keep the portrait within the mother country.[35] Now owned by a small gallery in Banks's home county of Lincolnshire, the portrait deemed too precious to go abroad is doubly hard to see: consigned to a remote part of England, it hangs high up and poorly lit on the wall of a staircase.

· CHAPTER 4 ·

EXPLORATION AND EXPLOITATION

... expeditions may fail in the main object of the arduous enterprise; but they can scarcely fail in being the means of extending the sphere of human knowledge ... 'Knowledge is power'.
John Barrow, Quarterly Review, 1818

In 1996, a Canberra art exhibition called *The Clever Country* commemorated Australia's scientific pioneers. Although the organisers wanted to stress local originality and national independence from Britain, several pictures celebrated Joseph Banks, the Englishman who had spent a few weeks collecting biological specimens to take back home and later converted the far-flung colony into a convenient dumping ground for unwanted criminals.

Among the first exhibits was a large 18th-century canvas (Figure 9). Now in Australia's National Library, this picture was by an English artist who specialised in conversation pieces, those fashion-able group portraits designed to show off the possessions and behaviour of wealthy families and their friends as they engaged in polite chit-

Figure 9. 'James Cook, Joseph Banks, the Earl of Sandwich and two gentlemen' (1777), by John Hamilton Mortimer. (By permission of the National Library of Australia.)

chat. Here these five gentlemen with their faithful dogs curled up at their feet have been displaced from their London drawing-rooms. Instead, they pose elegantly on the Yorkshire coast at the country estate of Banks's friend Constantine Phipps.

In the centre stands James Cook, wearing his naval uniform and holding the traditional gentle-manly pose of the Apollo Belvedere statue. Turning to the finely-dressed Earl of Sandwich, First Lord of the Admiralty, Cook gestures with his outflung hat to indicate that he has crossed – perhaps conquered – the seas, and may soon set out on another voyage. Banks, perched on a conveniently chair-shaped rock, sports a striking ensemble in red, white and blue, and completes the triangle of British naviga-tors who visually tie the scene together.

This picture was prominently displayed in *The Clever Country*'s galleries and reproduced on the first page of the catalogue. Many visitors must mistakenly have imagined that it showed Australia: even the vegetation provides little guidance, since the earliest artists anglicised the appearance of the alien Pacific landscape. This picture invites rein-terpretation as an allegory of imperial dominion and the conversion of Australia into a distant part of Britain. On this reading, with his right hand Cook proclaims ownership of the Pacific Ocean; the papers in his left hand could be either his new navigational chart or his orders to capture foreign territory for Britain. Sandwich leans nonchalantly against a classical statue of a half-naked woman, who might symbolise virgin territory ready to be overpowered. Although Sandwich had not actually travelled on the *Endeavour*, the voyage depended on his patronage. Gazing up towards him, Banks proffers a manuscript, possibly drawings to emphasise that his botanical studies could make an important contribution to science and naval exploration.

Modern Australians may want to take Banks over as *The Clever Country*'s first scientist, but in fact Banks helped to take over Australia for Britain. Cook was highly acclaimed for using new scientific instruments to chart the Pacific more accurately, but his primary allegiance was to Sandwich and the Admiralty rather than to the Royal Society. Naming is a statement of possession – so when Wallis first landed in Tahiti, he called it King George's Island. Botany Bay, where Cook's two botanists collected an astonishingly large number of previously unknown plants, is bordered by Point Solander and Cape Banks. When Banks gave Pacific plants a Linnaean name, he made them part of European science but suppressed their local identity – an entire genus of Australian shrubs and trees is called *Banksia* under the Linnaean system.

Was this a scientific expedition or a journey of imperial exploration? The answer, of course, is both. It is impossible to disentangle botanical and astronomical discoveries from commercial and colonial expansion. Banks was a self-funded researcher who later controlled the development of Britain's growing empire; he had been granted permission to travel with Cook on a voyage that was ostensibly about astronomy, but was in reality a naval mission. As in so many research projects that supposedly pursue pure knowledge, science and the state were inextricably tangled together.

* * *

In June 1760, the Fellows of the Royal Society were dismayed to learn that Britain's traditional enemies, the French, had already organised several expeditions to record the transit of Venus the following year. Emphasising that national honour was at stake, they promptly sent off a letter demanding £800 from the British government. Surely, this letter argued, 'it might afford too just ground to Foreigners for reproaching this Nation [if] England should neglect to Send observers to Such places as are most proper for that purpose and Subject to the Crown of Great Britain'.[36] So, to maintain the national reputation, two overseas expeditions were funded. The Fellows were probably secretly delighted when they learnt that one of the astronomers on the French project had been foiled because British forces captured his intended observation base in India. But although several countries participated in the project, the results were inconclusive, and the Royal Society resolved to be better prepared for the next transit, due to take place in 1769.

Astronomers wanted to observe Venus (or other planets) passing in front of the Sun in order to determine the mean distance between the Earth and the Sun, which is a fundamental unit used for calculating the size of the universe. An accurate result entailed comparing measurements of the time Venus took to cross the Sun's disc as seen from many different places in the world. The 1769 transit was a perfect opportunity for putting into action that worthy if unrealistic scientific ideal – international cooperation. Rather than setting up joint experiments, separate national teams later exchanged readings. To avoid a repetition of the 1761 embarrassment, this time the British were determined to lead the way. The Royal Society demanded and got £4,000 from the King to sponsor four expeditions, one of them to Tahiti.

Much of the Tahitian money was spent on equipment. Cook's ship was supplied with modern brass instruments, including several telescopes, a barometer and three clocks, some of them custom-made for the voyage. One of the Fellows of the Royal Society designed a portable observatory which Cook's crew could assemble in Tahiti. Sailing long distances was dangerous in the 18th century – as Samuel Johnson quipped, being in a ship was like being in a jail with the added likelihood of getting drowned. So the *Endeavour* needed not only the astronomical instruments needed for measuring the transit of Venus, but also navigational tools for charting the ship's course across unfamiliar oceans and for mapping remote coastlines more accurately. In addition, Cook had to carry out scientific trials: the Royal Society gave him some new inventions to test, such as a redesigned magnetic compass, while the Admiralty asked him to try out different diets for warding off scurvy.

The Admiralty and the East India Company soon realised the benefits of combining an expedition to Tahiti with a reconnaissance mission to the South Pacific. For a long time, world maps had shown an 'Unknown Southland', often imaginatively combined with Marco Polo's enticing descriptions of Locac, a country rich in timber, gold and elephants. During the 17th century, Dutch explorers had sketched in some of Australia's coastline, but the French and the British governments were keen to tap the resources of this still mysterious continent. For once, Britain and France were officially at peace after the Seven Years' War, but both countries realised that controlling the Pacific zone was vital for defending colonial possessions and protecting trade routes.

The Royal Navy was involved from the early stages of planning, and it was the Royal Navy, not the Royal Society, which ensured that Cook – an experienced Navy man – was appointed as commander. Before the *Endeavour* set sail, Cook received secret instructions from the Admiralty which make clear that this was a government expedition to acquire new territories. Discovery and exploration, these extra orders explained, 'will redound greatly to the Honour of this Nation as a Maritime Power, as well as to the Dignity of the Crown of Great Britain, and may tend greatly to the advancement of the Trade and Navigation thereof'.

When Wallis had been in Tahiti, he had seen some mountains in the distance. Could these be part of the southern continent? As soon as Cook had completed his transit of Venus work, the Admiralty letter continued, he was 'in Pursuance of His Majesty's Pleasure hereby requir'd and directed to … proceed to the southward in order to make discovery of the Continent'. Bound to secrecy, Cook was told to collect information, claim land in the name of the British King, and send in all the ships' log-books – sealed – to the Admiralty at the end of his journey.

By the time that marines and artillery had been taken on board, there were, Cook recorded in his journal, '94 persons including Officers, Seamen Gentlemen and their servants, near 18 months provisions, 10 carriage guns 12 swivels with a good store of Ammunition and stores of all kinds'.[37] There were also some paying passengers – Banks and his retinue of companions. These included Solander and another Linnaean disciple, two black servants and the trained artist Sydney Parkinson, along with two dogs and a famous goat that had already been round the world with Wallis. (The black servants froze to death in Tierra del Fuego after collapsing in the snow, and one dog had a fatal seizure just before the *Endeavour*'s return; but the goat survived and – wearing a silver necklet engraved with a Latin couplet by Johnson – became a tourist attraction in Greenwich Park.) Banks's men may well have been closely inspected before they went on board, since during the earlier French expedition to Tahiti, the islanders (not the French) had discovered that the botanist's assistant was a woman disguised in men's clothes.

This was a large-scale naval operation – the *Endeavour* set off with 7,860 pounds of sauerkraut (part of the scurvy trials) and at Madeira, an extra 3,000 gallons of wine were taken on board. The ship was overflowing with people, equipment and food supplies, which included dried soup and carrot marmalade as well as pigs, sheep and chickens kept alive (until needed) in pens on the deck. The standard hammock allocation was fourteen inches per person, and Cook was obliged to share his small private cabin not only with Banks and Solander, but also with the specimens they were drawing.

Although Banks's official permission to join the *Endeavour* had arrived only at the last moment, he had evidently been planning the trip for several months. In addition to 30 large boxes, Banks's storage equipment included casks holding preserving liquids and over 200 bottles – to say nothing of the telescopes, microscopes and twenty-odd guns (with around 300 pounds of ammunition). Another botanist wrote to Linnaeus about Banks's luggage, which was rumoured to have cost £10,000. 'No people ever went to sea better fitted out for the purpose of Natural History', he marvelled – 'They have got a fine Library [and] all sorts of machines for catching and preserving insects; all kinds of nets, trawls, drags and hooks for coral fishing, they have even a curious contrivance of a telescope, by which, put into the water, you can see the bottom at a great depth.' And, he concluded unctuously to gratify Linnaeus, 'All this is owing to you and your writings.'[38]

Banks's collection grew rapidly, and ate into the remaining space. At every opportunity, the naturalists trawled for fish, retrieved birds and insects caught in the ship's rigging, and hunted for plants and animals when they went on shore. The naturalists settled into a routine which they maintained throughout the voyage – Banks and Solander inspected the new specimen, Parkinson or one of the assistants drew it, and then its new name and details were added to their Linnaean textbooks. Banks brought back around 3,000 dried plants and almost 1,000 original drawings, many of them made at sea from specimens piled up beneath damp cloths to keep them fresh. Cook was impressed by Banks's dedication, but some of the crew seem to have had reservations: when they set out to catch turtles on the Great Barrier Reef, one sailor engaged in what Banks called 'unaccountable conduct' that effectively stymied their chances of loading heavy, slippery turtles into the rowing boat.

From the time they left Plymouth on 25 August 1768, Banks kept a daily diary during the three-year voyage – over 1,000 entries that log his slow progress across the ocean until the *Endeavour* landed in Kent on 12 July 1771. Banks was far more interested in watching the outside world than in recording his inner experiences or his relationships with the men squashed in around him. This early entry gives a good indication of his determination to be a good Linnaean botanist as well as indirectly revealing some of the tensions that arose in the cramped conditions on board the *Endeavour*: 'About noon a young shark was seen from the Cabbin windows following the ship, who immediately took a bait and was caught on board; he proved to be the *Squalus Charcharias* of Linn and assisted us in clearing up much confusion which almost all authors had made about that species; with him came on board 4 sucking fish, *echineis remora* Linn. who were preserved in spirit. Notwithstanding it was twelve O'Clock before the shark was taken, we made shift to have a part of him stewd for dinner, and very good meat he was, at least in the opinion of Dr Solander and myself, tho some of the Seamen did not seem to be fond of him, probably from some prejudice founded on the species sometimes feeding on human flesh.'[39]

Despite Banks's apparently naïve enthusiasm for the joys of scientific research, he was forced to follow a naval regime and to recognise political realities. Cook ran his ship to a strict timetable punctuated by nautical piping to announce changings of the watch, taking meals and crawling into the hammocks for the night. At noon every day there was a special ceremony when the officers measured the position (altitude) of the Sun. The reading was ritually conveyed from person to person until it reached the captain, who formally announced the beginning of the nautical day, twelve hours out of kilter with civilians (which explains discrepancies between journals kept by Banks and Cook).

Banks could spend time ashore only when it suited Cook's schedule, and even then he was constrained by international relations. When they arrived at Madeira, the English consul arranged permits for them to travel round the island and collect plants, although – much to Banks's disgust – they had to waste a whole day of their brief visit being polite to the governor. At Rio de Janeiro, the governor thought 'it impossible that the King of England could be such a fool as to fitt out a ship merely to observe the transit of Venus', and was convinced that they were either smugglers or spies. Portuguese guards watched the ship closely, and although Banks sent begging letters, he was confined on board for three weeks even though the *Endeavour* was tilted over so that its sides could be cleaned. 'You have heard of Tantalus in hell', moaned Banks, 'you have heard of the French man laying swaddled in linnen between two of his Mistresses both naked using every possible means to excite desire but you have never heard of a tantalized wretch who has born his situation with less patience than I have done mine I have cursd swore ravd stampd'. He did, however, indulge in some illegal botanising by secretly climbing down a rope into a small boat during the night.[40]

By New Year they were down in the South Atlantic. Wrapped up in layers of flannel clothes, Banks happily observed species of birds and sea creatures that he had never encountered before. Sustained by rum and roast vulture, he led his team on a disastrous overnight expedition in Tierra del

Fuego, when he only just survived a bitterly cold snow storm. To Cook's astonishment, Banks and Solander went back later that day to collect some more shells and plants, but then they were obliged to sail onwards. Over 4,000 miles of uncharted waters lay between the *Endeavour* and Tahiti. Astronomy, not botany, was paying for this part of the journey, and Cook wanted to arrive several weeks before the transit on 3 June so that he could get his observatory ready.

* * *

In Banks's view, Tahiti belonged to Britain. This is how he recorded their arrival: 'This morn early came to anchor in Port Royal bay King George the thirds Island.'[41] Foreign ships had turned up before, and the Tahitians knew from experience that Europeans came equipped with guns – and were not afraid to kill. Very sensibly, they sent out canoes full of food and persuaded Cook to take part in a peacemaking ritual as soon as he landed with an advance party. This diplomatic welcome helped to convince the *Endeavour* travellers that they had landed among a peaceful, harmonious society.

The violence came primarily from the English side. Banks demonstrated the power of guns by killing three ducks with one shot, and within a few days the crew had killed an islander who took a sentry's musket. Cook tried hard to keep his sailors under control, but there were frequent disputes during the three months they stayed on the island. When the ship's butcher snatched a stone axe and threatened its owner with a reaping-hook, Cook had him flogged. Banks compared English discipline with Tahitian sensitivity: 'they stood quietly and saw him stripd and fastned to the rigging but as soon as the first blow was given interfered with many tears, begging the punishment might cease a request which the Cap^{tn} would not comply with.'[42]

In spite of the overt hospitality displayed by the islanders, Cook decided that the British had to be barricaded in for their own safety. Enrolled to chop and carry wood, the Tahitians helped to build the military garrison that was designed expressly to keep them out. Plagued by flies and blowing sand, Parkinson temporarily abandoned his flower drawings to sketch Venus Fort (Figure 10), the first European settlement in the Pacific, which was patrolled by sentries and protected by several large guns on the ramparts. Nevertheless, Parkinson portrayed a tranquil scene. As the British flag flutters in the breeze, smoke from the oven rises above the protective moat and palisades, while the local men in their boats may well be fishing to feed their uninvited guests.

The Tahitians were mystified by some of the

Figure 10. 'Venus Fort' (1769), by Sydney Parkinson. (By permission of the Syndics of Cambridge University Library.)

bizarre behaviour they witnessed – saluting a piece of material flapping at the top of a tall pole, parading up and down with a drummer, wearing heavy clothes completely unsuited to the climate, looking through brass tubes at the stars ... They strategically placated their dangerous guests by supplying them with food, but the visitors offered only a mean rate of exchange. Realising that their whole island was being taken over by the unexpected arrivals, the islanders appropriated some snuff-boxes, nails and magnifying glasses. Ignoring their own discourteous behaviour which breached local etiquette, the Europeans repeatedly accused the Tahitians of theft.

For the islanders, taking strange items from the Europeans was risky because of their guns, but did promise excellent bargaining opportunities. Only a few weeks before the transit, a treasured possession disappeared – the purpose-built quadrant designed to measure astronomical angles with unprecedented precision. Without it, the whole expedition was meaningless. A local informant seized the opportunity to gain a reward, and guided Banks as he chased after the missing instrument in the stifling heat. Although he eventually retrieved the quadrant, the Tahitians perhaps enjoyed watching Banks's terror as he contemplated being 'at least 7 miles from our fort where the Indians might not be quite so submissive as at home'.[43]

While Cook prepared for the transit, Banks spent his time collecting plants and enjoying himself with the Tahitians, whose clothes and customs he recorded with the same meticulous attention to detail that he paid his scientific specimens. Although Banks did not realise it at the time, his interaction with the local people was exceptionally intimate here. Afraid of the European guns and used to trading with other Polynesians, the islanders allowed Banks to participate in their dances and ceremonies, and exchanged information about how crops should be cultivated and how local plants could be used for food and medicine. The curiosity was mutual – two men even tried out the other group's shaving techniques.

Because Cook was worried about clouds obscuring the sun at the vital moment, he sent Banks with a couple of astronomers to watch the transit from a nearby island. This short expedition proved a great success for everyone. The weather was perfect, so the observers were pleased, while Banks found some new plants as well as '3 hansome girls'. They presumably thought the adventure would prove profitable, since they 'with very little perswasion agreed to send away their carriage and sleep in the tent, a proof of confidence which I have not before met with upon so short an acquaintance'.[44]

* * *

After three months, it was time to leave Tahiti: the astronomical measurements had been completed, the Europeans had more or less exhausted the local food supplies, the islanders increasingly resented the vindictive punishments meted out by their visitors, and some of the sailors were plotting to mutiny and remain behind. Obeying his secret instructions from the Admiralty, Cook set off south-wards to look for Australia. Fortunately for the *Endeavour*, Banks paid for his friend Tupaia, a high-ranking priest, to accompany them with his young son Tayeto. Banks reflected that he was rich enough 'to keep him as a curiosity, as well as some of my neighbours do lions and tygers', and anticipated with delight 'the amusement I shall have in his future conversation'.[45] Without this hired 'curiosity', Cook and Banks might well not have survived. Tupaia's expert knowledge of the currents, islands and local languages rescued them from several sticky situations, and he showed them how to find and cook their food.

Weeks later, because there was still no sign of Australia, Cook headed towards the certainty of New Zealand, which had been mapped by Dutch explorers over a hundred years earlier. As they neared land at last, excitement mounted in the cramped quarters. If only, mused Banks, our friends in England could see us now: 'D[r] Solander setts at the Cabbin table describing, myself at my Bureau Journalizing, between us hangs a large bunch of sea weed, upon the table lays the wood and barnacles; they would see that notwithstanding our different occupations our lips move very often, and without being conjurors might guess that we were talking about what we should see upon the land which there is now no doubt we shall see very soon.'[46]

In October 1769, almost three months after leaving Tahiti, the *Endeavour* arrived at New Zealand and eventually sailed right round both islands in a figure of eight (hence the name Cook Strait between them). Although they landed several times, in comparison with Tahiti these stops were fraught with conflicting interests. One problem was deciding where to go: Cook was trying to find a suitable site for observing the transit of Mercury, while Banks's major concern was to collect plants. In addition, the Maoris wanted more information about these foreign arrivals. Gathering together, they politely raised a single spear and chanted so that the visitors would declare their intentions. Misinterpreting this invitation as hostility, the Europeans fired back with their guns. For Banks, there was only one way to deal with people he regarded as cannibals: 'They always strenuously oppos'd us so that we sometimes were laid under the disagreeable necessity of effecting our Landing by Force. They were, however, when subdued, unalterably our friends.'[47]

To supplement their dwindling food supplies, Banks searched for edible vegetables, noting for the benefit of future travellers that plenty were available in the autumn. Among the 400 plant species they discovered, Banks was most impressed by the flax, used locally to make clothes and fishing nets. Although the Maori called their plant Harakeke, Banks gave it a Linnaean label – *Phormium tenax* (he even christened one variety *Phormium cookii*). He gloated that 'so usefull a plant would doubtless be a great acquisition to England', and Parkinson carefully drew flax for the book they planned to publish together on their return (see Figure 6).[48]

After more than a year at sea, Cook was worried about the *Endeavour*'s condition, but still hoped to find new territory. A few weeks after they set off from Cape Farewell (as it became known on European maps) they came across an unexpected coast, and sailed northwards along it trying to land. At first Banks was not impressed: 'The countrey … resembled in my imagination the back of a lean Cow, coverd in general with long hair, but nevertheless where her scraggy hip bones have stuck out farther than they ought accidental rubbs and knocks have intirely bard them of their share of covering.'[49] At last they managed to land in Botany Bay: they had stumbled on Australia by accident.

The aboriginal people prudently abandoned their weapons after they had learnt what guns could do; from behind the safety of his pistols, Banks boasted that he felt 'quite void of fear as our neighbours have turnd out such rank cowards'. The Europeans raided the local settlements, taking away interesting objects as well as eating the food that was cooking on the fires. They left behind some trinkets as gifts, but the recipients did not always appreciate these cheap examples of superior British civilisation – during his botanical forays, Banks found them piled up and abandoned.

Banks was overwhelmed by the wealth of new discoveries he made, spending days drying out his plants on a sail in the sunshine. Dining off huge stingrays and bustards, Banks, Solander and Parkinson desperately tried to catalogue and draw all their specimens, while Cook carefully mapped the coastline, giving prominent features English names such as Cape Sandwich (after the head of the Admiralty that was funding the trip).[50]

As they continued sailing towards the north, they encountered the Great Barrier Reef – another accidental discovery made when they ran aground on the submerged rocks. It took weeks to repair the large hole in the *Endeavour*. Banks was in despair: 'Since the ship has been hauld ashore the water that has come into her has of course all gone backwards and my plants which were for safety stowd in the bread room were this day found under water; nobody had warnd me of this danger which had never once enterd into my head; the mischeif was however now done so I set to work to remedy it to the best of my power. The day was scarce long enough to get them all shifted &c: many were savd but some intirely lost and spoild.'[51]

Over 200 years later, we know that this is a story with a happy ending, but the travellers themselves had no such security. As the Europeans leisurely mended their ship, they outstayed their welcome. The local inhabitants were appalled by the behaviour of their visitors, who refused to share the turtles they had caught for dinner and hunted for the nearest village so that 'we might have an opportunity of seeing their Women'. Infuriated, they interrupted one of Banks's plant-gathering expeditions by setting fire to the grass, and so successfully forced the invaders to leave. But first the Europeans had to negotiate their way through the treacherous uncharted reef. As the breakers crashed round the fragile ship, Banks forgot to worry about his precious collection: 'The fear of Death is Bitter', he told his journal; 'the prospect we now had before us of saving our lives tho at the expence of every thing we had made my heart set much lighter on its throne.'[52]

A couple of weeks later, they sailed past the northern tip of Australia. A small party rowed ashore and, hoisting the British flag, Cook announced to the empty landscape that King George III now owned the land to the south – and so New South Wales became marked out on the map as a British possession. After a brief stay in Papua, where the guns came in handy again, they headed for England, stopping off from time to time to load up with provisions.

Their mission was completed, but they were still a year away from England. By now, the *Endeavour* was limping, everyone except Banks and Solander was feeling homesick, and fever was decimating the travellers – Tupaia and Tayeto, Parkinson, the astronomer, the cook, the surgeon and many others died along the way. Banks was ill for weeks, but eventually he arrived safely in London where he was immediately summoned to court so that George III could hear about his adventures.

* * *

The King was not a seafaring man: once when he inspected a warship, he was unsure whether to go down a ladder backwards or forwards. He was, however, fascinated by agricultural reform, and he was far more interested in the potential of Banks's botanical discoveries than in talking to Cook. It was Banks rather than Cook who returned a hero, and the flurry of caricatures and satirical poems indicate how rapidly he became prominent among London's gossipy social élite.

Once back in his own home, Banks started to unpack. 'His house is a perfect museum', exclaimed an awestruck visitor, who wandered through rooms stuffed with weapons, clothes and ornaments before admiring the arrays of animals preserved in spirits and 'the choicest collections of drawings in Natural History that perhaps ever enriched any cabinet, public or private: – 97 plants drawn and coloured by Parkinson; and 1300 or 1400 more drawn ... and what is more extraordinary still, all the new genera and species contained in this vast collection are accurately described, the descriptions fairly transcribed and fit to be put to the press.'[53]

This was an optimistic assessment. Although twelve years later, Banks confidently announced that he had just a couple of months' work left, he never did complete his ambitious publishing project – a mammoth illustrated *Florilegium* with 743 plates illustrating all the new plants he had discovered. Despite the help of Solander and other assistants, as well as his sister Sarah (who thoughtfully cleaned up his grammar), Banks never even got round to publishing his journal. Why did he keep postponing this work? Several explanations have been put forward – Solander's death, quarrels with Parkinson's relatives, Banks's unease with his own writing ability, the birth of his illegitimate child, pique at not sailing with Cook again … Perhaps he simply became so immersed in other projects that he kept putting things off (most people are familiar with the *mañana* syndrome).

The 1885 *Dictionary of National Biography* sneered that Banks's 'writings are comparatively trifling'. This is true – his most important published works were specialised pamphlets on wool and corn blight. Banks's scholarly reputation would undoubtedly have been far higher if his publication record had also included the *Florilegium*. But books are not necessarily the best way of measuring scientific achievement. As President of the Royal Society, Banks wielded enormous influence and initiated changes that had permanent effects.

For one thing, Banks made Linnaean botany central to British science. He was justified in boasting towards the end of his life: 'How immense has been the improvement of botany since I attached myself to the study, and what immense facilities are now offered to students, that had not an existence till lately!'[54] Still more significantly, by demonstrating how useful foreign plants could be, Banks reinforced the links between commercial, imperial and scientific exploration. He sponsored further overseas research, and the Admiralty started regularly including a naturalist on its expeditions. And that was why Charles Darwin came to travel with HMS *Beagle* and observe the plants and animals which were crucial for his theory of evolution by natural selection. For half a century, Banks dedicated his life to making science work for the state – and making the state pay for science.

· Chapter 5 ·

Exoticism And Eroticism

I am apt to suspect the negroes and in general all other species of men (for there are four or five different kinds) to be naturally inferior to the whites. There never was a civilized nation of any other complexion than white, nor even any individual eminent in either action or specula- tion ... On the other hand, the most rude and barbarous of the white, such as the ancient Germans, the present Tartars, have still some- thing eminent about them, in their valour, form of government, or some other particular. Such a uniform and constant difference could not happen, in so many countries and ages if nature had not made an original distinction between these breeds of men.

David Hume, Of national characters, 1754

For Joseph Banks's friends James Boswell and Samuel Johnson, travelling to the Hebrides was like going abroad. Surrounded by people who spoke no English, Boswell crouched on a grass seat and gazed at the exotic scene in front of him. 'It was much the same as being with a tribe of Indians', he remarked to Johnson. 'Some were as black and wild in their appearance as any American savages whatever. One woman was as comely almost as the figure of Sappho.'[55] Like Linnaeus in Lapland and Banks in Tahiti, Boswell seems torn between emphasising his bravery as a witness and recording his hosts' behaviour with anthropological detachment.

As British gentlemen, Boswell, Johnson and Banks believed that they were superior to the foreigners they met. This confidence stemmed from two major sources: the Bible and Aristotle. By reading the book of Genesis, Christians learnt that God had created human beings separately, giving them the privilege as well as the responsibility of looking after His Earth and using it for their own benefit. Aristotle had envisaged the natural world arranged in order along a great chain of being, starting with rocks and the humblest organisms at the bottom, and gradually moving upwards through plants, fish and animals to reach human beings. White European men were, of course, right at the top of this ladder.

Well into the 18th century, many naturalists still thought that this chain never changed, so that the present world of living creatures was exactly the same as the one that God had originally created. But particularly after explorers brought back new species from America, it became increasingly difficult to squeeze every living organism into a single straight line with only minute changes between one rung of the ladder and the next. All sorts of problems arose. Should cats be higher than dogs? Where should reptiles go – above fish or below them? What about whales? And did it really make sense to put stones right next to moulds and lichens? Natural philosophers started tinkering with the chain, proposing branching systems more like trees than ladders, and suggesting that perhaps the Earth's living occupants had developed over a period of time.

When Europeans travelled overseas, they encountered societies very different from their own. Here was yet another dilemma of classification. Should these people be ranked in order below Europeans as part of the continuous chain, or should all human beings be split off together into a separate group of their own? Both solutions raised problems. Human beings are special because they have a soul – but do animals have souls? Can people be moved up the rungs of the ladder by exposing them to European civilisation? Who should be placed higher, European women or Asian men? Where should the boundary be drawn between naked primitive savages (their vocabulary, not mine) and very intelligent apes? And so on.

During the 18th century, natural philosophers adopted three main approaches for explaining the variations between human beings. The first two –

often labelled climatic and subsistence theories – emphasised the ways that people lived; the third – the taxonomic method – was initiated by Linnaeus and relied on classifying humans by their appearance and behaviour.

Climatic accounts were the oldest. Since the time of Hippocrates, writers had attributed the differences between people to environmental conditions. For instance, they explained that the sun scorched Africans dark and made them lethargic – in any case, they had little incentive to work because the constant sunny weather made the land so fertile. Enlightenment natural philosophers found such climatic explanations appealing because they were compatible with the biblical account of creation. After some sort of Fall, the various versions of these theories agreed, the members of a single original group of people had dispersed to different parts of the world and then adapted themselves to the local environment.

The most influential advocate of climatic theories was the Comte de Buffon, who published an enormously successful multi-volume study of natural history in the middle of the 18th century. Originally appearing in France, it was almost immediately translated into English and became famous all over Europe. Although Buffon was one of Linnaeus's major critics, the two naturalists did share a faith in their own superiority. Firmly separating humans from other animals, Buffon divided them into two major categories, typified by 'our great civilised peoples' and 'the little savage nations of America'. On Buffon's account, the Arctic climate had ruined the character as well as the physique of the Sami: 'the women are as ugly as the men, and indeed resemble them so strongly that one cannot tell them apart ... they are coarser than savages, without courage or self-respect or modesty; this abject people has customs one can only despise.'[56]

In contrast, the English philosopher John Locke had put forward a four-stage subsistence theory, which became particularly important in Scotland. These were progressive models which hinged on methods of finding food and owning land. Initially humans were hunters like carnivorous animals, but as they became more civilised they moved upwards through the next three stages. First herders domesticated animals, then farmers established permanent agricultural settlements, and finally – as in Western Europe – commercial organisations appeared. According to this scheme, societies stuck at earlier levels could be improved and brought up to Western standards – an attractive prospect for liberal educators and Christian missionaries. Moreover, subsistence theories implied that progress was possible for the whole of humanity. As the religious chemist Joseph Priestley explained, 'It is nothing but a superior knowledge of the laws of nature, that gives Europeans the advantages they have over the Hottentots ... science advancing, as it does, it may be taken for granted, that mankind some centuries hence will be as much superior to us ... as we are now to the Hottentots.'[57]

Writers did use the word 'race' in the 18th century, but not in the same way that we do now. The modern concept of race originated with Linnaeus, who introduced the taxonomic way of grouping people. For his modified version of the linear Aristotelian chain, Linnaeus drew a two-dimensional map. First he split the universe into three major kingdoms – minerals, plants and animals – and then subdivided each of these into orders, classes and so on. Although controversial, his innovations did bring great advantages. For instance, it no longer mattered whether the cat family was shown above or below the dog family, since their location on the page implied nothing about their position in a continuous hierarchy.

One feature of Linnaeus's system that Buffon and many other naturalists especially disliked was his way of classifying human beings. For one thing, they accused Linnaeus of being influenced by preconceived convictions rather than basing his ideas on observations. The Bible was particularly important for this Lutheran clergyman. Just as there were four rivers in the Garden of Eden and four continents, so too Linnaeus decided that there must be four human races. This also corresponded to Aristotelian ideas that the universe is composed of four elements – earth, air, fire and water – and that human health is governed by four humours. Unsurprisingly, Linnaeus's top race was *Europaeus albus*, the ingenious and sanguine white Europeans. The other three were the happy-go-lucky Red Indians, the melancholy yellow Asians, and the idle black Africans.

Although Linnaeus still put Europeans at the summit of creation, to his opponents' horror he placed people in the same order – *Anthropomorpha* (human-like creatures) – as apes (Figure 11). Justifying himself by emphasising physical resemblances, Linnaeus converted human beings into the close relatives of apes. 'No one has any right to be angry with me', he wrote touchily; 'as a natural historian according to the principles of science, up to the present time I have not been able to discover any character by which man can be distinguished from the ape.'[58] His four human races were subdivisions of the species *Homo sapiens* – wise man – but Linnaeus refused to place them in a separate category of their own. Instead, he suggested that other species of *Homo* also exist. Although he changed his ideas over time, in Figure 11 the two creatures on the right (Satyr and Pygmee) are apes behaving in a humanoid way, while the two on the left are both human women, classified as *Homo*. The

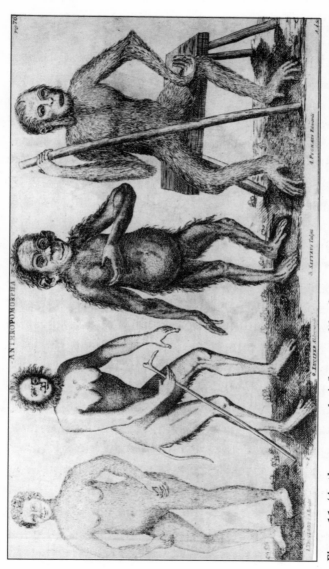

Figure 11. 'Anthropomorpha' (human-like figures). From Carl Linnaeus, 'Anthropomorpha', in *Amoenitates Academicae* (1764). (By permission of the Syndics of Cambridge University Library.)

hairier one, Lucifer, represents *Homo caudatus*, tailed man, while the other is *Homo troglodytes* – caveman or night person.

Never having seen an example of *Homo troglodytes* himself, Linnaeus had adapted this figure from a much older drawing of an orang-utan. However, he prudently left out the 'Hottentot apron' – the discreet label for elongated sexual organs that European explorers had observed among Hottentot women – although he did describe it in the text. Along with other naturalists, Linnaeus alleged that, because of their closeness to animals, all African women possessed these aprons. Obsessed with finding and measuring them, explorers debated whether they occurred naturally or had been stretched artificially to comply with the Hottentot sense of fashion. One French explorer modestly (if unconvincingly) explained that he had been reluctant to make his shy Hottentot informant undress, but had persevered with this intimate examination in the interests of scientific research.

As natural philosophers argued about the differences between animals and people, they often focused on speech. A French cardinal tried to prove his point by addressing an orang-utan in the zoo. 'Speak, and I will baptise you', he declared – but even that theatrical test was not enough to resolve the puzzle.[59] There seemed to be no hard boundary: parrots could be trained to talk, and travellers brought back local rumours that apes could also speak. In spite of Buffon's contempt, Linnaeus insisted that his *Homo troglodytes* communicated by a guttural hissing language that was too difficult even for Europeans to learn.

And there was another problem – the wild children, those boys and girls who had been discovered living with and like animals. These children became great curiosities, sources of lurid fascination who attracted experts from all over Europe to gaze at them and pass judgement on whether they were animals or people. As well as being speechless, they were often reported to shuffle on all fours, to eat raw food and to have an uncommonly keen sense of smell – one of them was even rumoured to have started life as a bear cub. Especially in France, they became the subjects of living experiments to determine where the boundaries should be drawn between human and beast. Linnaeus placed them in yet another species of people, *Homo ferus* or wild man, broadly characterised as 'four-footed, mute, hairy' and subdivided to accommodate individual children as they were retrieved.[60]

As a Bible-loving pastor, Linnaeus did not believe in evolution. According to him, his *Anthropomorpha* were not extinct but thrived in remote areas of the world. Buffon sneered that he had been misled by travellers' tales of orang-utans or albino Africans, but Linnaeus insisted on their reality. Although it might seem extraordinary that Linnaeus could sincerely believe in *Homo troglodytes*, travellers often gave distorted accounts of strange animals they had glimpsed in the distance. Some rumours of exotic humanoid creatures persisted for centuries – mermaids, for instance, or the giants of Patagonia (southern Argentina) originally reported by Portuguese mariners.

Linnaeus placed the Patagonian giants in the species *Homo monstrosus* (which also included Alpine dwarfs). The year before Banks sailed off in the *Endeavour*, accounts had been read out at the Royal Society from British explorers who confirmed that they had seen these monstrous South American people with their own eyes. The frontispiece of one navigator's travel account looks suitable for illustrating the satirical novel *Gulliver's Travels*: it shows a sailor timidly offering a biscuit to a woman almost twice his height, clothed in furs and carrying an enormous baby. Could these giants be real? Voyagers swore that 'there was hardly a man there less than eight feet, most of them considerably more', but satirists had a field day. We should seize this marvellous country for Britain, one of them declared – their giant trees would make marvellous ships, their gold and diamonds must be incredibly valuable, and their women 'could mend our breed, which, all good Patriots assert, has been dwindling for some hundreds of years'.[61]

Naturalists were divided, but like Buffon and Linnaeus, they were conducting this argument from the comfort of their European studies. Armchair taxonomists depended on Banks and other explorers for first-hand accurate descriptions not only of strange plants and animals, but also of the people whom they encountered during their voyages overseas. Although the *Endeavour* sailed past the Patagonian mainland without landing, Cook did let Banks spend a few days exploring the island of Tierra del Fuego off the tip of South America. Keen to play his part in settling the controversy about giants, Banks carefully noted that the local inhabitants 'are of a reddish Colour nearly resembling that of rusty iron mixd with oil: the men large built but very clumsey, their hight from 5f'8 to 5f'10 nearly and all very much the same size, the women are much smaller, seldom exceeding 5f''.[62]

But even such apparently precise observations (did he really persuade these people to stand against a tree so that he could mark their heights like growing children?) were not enough to settle the arguments. One obvious rejoinder was to claim that this negative evidence did not necessarily disprove the claims of earlier explorers. Since Banks had seen only about 50 people, it was easy to argue that the giants lived in another area or had hidden from the inquisitive invaders carrying their guns. Convinced since childhood that Europe was a unique haven of civilised normality, British people were reluctant to relinquish their beliefs that remote lands were teeming with bizarre occupants, and that the laws of nature were different in the southern hemisphere from in the north.

* * *

Through his voyage on the *Endeavour* and his subsequent work back in England, Banks did much to break down such prejudices. Although explorers brought back thousands of exotic specimens, they found no trace of Linnaeus's troglodytes or hairy humanoid monsters. Furthermore, now that Australia had been discovered, Linnaeus's insistence on a four-continent system no longer made sense. Banks's observations proved vital in disproving Linnaeus's four-fold classification of humanity. He helped to consolidate the nascent science of anthropology by bringing back skulls so that naturalists could study the inhabitants of far-flung lands without leaving their European laboratories. By carefully recording the societies he visited, Banks and his colleagues provided convincing evidence of the similarity between human beings throughout the world. Although Linnaeus's name *Homo sapiens* was preserved, it became a single order with only one species – in other words, humans were placed in a special category of their own.

From his secret Admiralty instructions, Cook knew that he should colonise land wherever possible. The Royal Society had also provided Cook, Banks and Solander with guidelines on how to deal with the territory occupied by indigenous people. This advice began tolerantly, pointing out that 'natives' are human beings, even if not quite at the same level as English gentlemen. 'No European Nation', the travellers were told, 'has a right to occupy any part of their country, or settle among them without their voluntary consent'. However, then followed some useful tips on how to obtain this 'voluntary consent'. Although killing was not recommended, other ways 'to convince them of the Superiority of Europeans' included firing a bullet through a hut, giving them mirrors, and pantomiming thirst and hunger so that even the 'most stupid … must immediately comprehend'.[63]

Banks set out to record the appearance and behaviour of the people he met with the same meticulous attention to detail that he gave his plants. Much of the time Banks commented on his hosts' appearance with what he perhaps thought was scientific detachment, but which often appears more like insulting condescension. Among the Tahitians, he wrote, are 'some handsome men and women, the only bad feature they have is their noses which are in general flat, but to balance this their teeth are almost without exception even and white to perfection, and the eyes of the women especialy are full of expression and fire'.[64]

In other places, Banks whetted his readers' appetites by telling them in advance what their moral judgement should be. 'One amusement more I must mention tho I confess I hardly dare touch upon it as it is founded upon a custom so devilish, inhuman and contrary to the first principles of human nature that tho the natives have repeatedly told it to me ... I can hardly bring myself to believe it much less expect that any body Else shall.' After this enticing introduction, he revealed 'that more than half of the better sort of the inhabitants of the Island have like Comus in Milton enterd into a resolution of enjoying free liberty in love ... seldom cohabiting together more than one or two days'. Although a few lines later Banks admitted that he had never himself witnessed one of those private sessions in 'which they give full liberty to their desires', English people readily believed his accounts of free sex, erotic dances and frequent abortions.[65]

Like Linnaeus, Banks automatically classified everything, and for Englishmen, sexual promiscuity implied being placed lower down the chain of being. Banks not only ranked the islanders below Europeans, but also set up an internal Tahitian hierarchy based on sexual activity: 'The men as I have before said are rather large, I have measured one 6 feet 3½; the superior women are also as large as Europæans but the inferior sort generaly small, some very small owing possibly to their early amours which they are much more addicted to than their superiors.'[66]

Pacific islanders did not, of course, realise that they were supposed to be inherently inferior to Europeans, and they were as amazed by these encounters as the Europeans. Linnaeus had maintained that his troglodytes communicated by hissing, but for the New Zealanders, Cook's 'language was a hissing sound, and the words he spoke were not understood by us in the least'. From their perspective, the visitors with their peculiar white skin and blue eyes behaved very strangely. For one thing, they rowed their boats backwards as though they had eyes in the backs of their heads – could they be some sort of supernatural beings? Once on land, these odd arrivals climbed cliffs to gather grass and kept tapping the stones on the beach. Some Maoris tried to help these botanical and geological investigations by gathering up stones and plants, but they noticed in bewilderment that 'some of the stones they liked, and put them into their bags, the rest they threw away'.[67]

The Maoris appreciated their great advantage: the British travellers needed food. Figure 12, a watercolour painted while the *Endeavour* was in New Zealand, shows Banks bartering with a local

Figure 12. 'Joseph Banks bartering with a Maori'.
Anonymous watercolour (by Tupaia), New Zealand
(1769). (By permission of the British Library.)

trader, who seems – contrary to many British
accounts – not the slightest bit intimidated. In this
symmetrical image, both men are standing in the
same posture and are wearing their own national
costumes; even the quiff of hair on the Maori's
head is matched by the elegant tied wig beneath
Banks's naval hat. Portrayed at the very moment of
an equal interchange, the New Zealander is handing
over a large red crayfish while Banks proffers a piece
of white bark cloth from Tahiti, which was more
in demand than the English gifts carried on the

Endeavour (although the price plummeted after the first day). The unknown artist has depicted an encounter between two cultural ambassadors, each of whom is carrying out a transaction with an unfamiliar foreigner.

Unpublished until the late 20th century, this record of an early encounter between Europeans and their Pacific hosts is very different from the ones that appeared in England soon after the *Endeavour*'s return. Like other travel narratives, the published accounts of the *Endeavour* voyage were heavily coloured by preconceptions of how Pacific people should look and behave. Many of the original pictures were polished up to conform with Western stereotypical views of exotic foreigners. The editor of Banks's and Cook's journals, John Hawkesworth, approached his task creatively, embroidering their narratives and imaginatively inserting anecdotes of indigenous Australians into stories about the Fuegians – after all, from the European perspective, they were all Pacific primitives and so were interchangeable.

Educated gentlemen like Banks and Linnaeus juggled apparently contradictory views of the remote regions they visited. On the one hand, they regarded Sami and Pacific islanders as inferior primitives who were dirty, uncivilised and far closer to animals than themselves. Transmitted through Hawkesworth, Banks's descriptions of ritual murders and other practices confirmed British superiority. But at the same time, Europeans admired such people for being noble savages, for living in an innocent, uncorrupted state, untainted by the depravity of modern civilisation and unburdened by the necessity of earning their living. In Tahiti, wrote Banks, 'Love is the Chief Occupation ... both the bodies and the souls of the women are modeled into the utmost perfection for that soft science idleness the father of Love reigns here in almost unmolested ease, while we inhabitants of a change-able climate are oblige to Plow, Sow, Harrow, reap, Thrash, Grind Knead and bake our daily bread.'[68]

Drawing on biblical imagery, Europeans referred to the Pacific area as though it were an earthly paradise existing before the Fall. The first French visitor to Tahiti 'thought I was transported into the garden of Eden; we crossed a turf, covered with fine fruit trees, and intersected by little rivulets ... everywhere we found hospitality, ease, innocent joy, and every appearance of happiness amongst them.' The ground beneath the trees was scattered with ripe breadfruit, a name recalling God's manna that fell from the sky.

Those with the benefit of a classical education also thought of the Pacific as Arcadia, the idyllic countryside of the Greek Golden Age where nymphs and satyrs idled away their days in romantic courtship. Banks remarked that, apart from their complexion, Tahitian women were far superior to the English beauties he had left behind because they were naturally elegant and wore loosely draped clothes like Greek goddesses. 'The scene that we saw', he enthused, 'was the truest picture of an Arcadia of which we were going to be kings that the imagination can form.' He even gave the Tahitian men Greek names, such as Hercules (for his strength) and Epicurus (for his appetite).[69]

Drawings that were published after the *Endeavour* voyage consolidated this vision of an Elysian paradise, but the travellers' immediate impressions were different. Cook thought the Fuegians were 'perhaps as miserable a set of People as are this day upon Earth', and reported that 'the Women wear a peice of skin over thier privey parts but the Men observe no such decency. Their Hutts are made like a behive and ... cover'd with branches of trees, long grass &c in such a manner that they are neither proff against wind, Hail, rain or snow.'[70] An on-the-spot sketch made by one of Banks's assistants confirms Cook's bleak description. Draped in rough skins, a small group of swarthy people huddle round some smouldering logs; squatting inside a dilapidated hut, they are accompanied by an animal resembling a large rat.

This image was heavily doctored for Hawkesworth's account of the voyage (Figure 13). Suspiciously English-looking trees now surround a

Figure 13. 'A View of the Indians of Tierra del Fuego in their Hut'. Engraving after Giovanni Cipriani by Francesco Bartolozzi (1773). (By permission of the Syndics of Cambridge University Library.)

cosy hut which is appealingly natural, but also well-maintained. Inside, the members of a happy family, which now includes chubby children, toast themselves before a blazing fire. These light-skinned people are elegantly dressed in Grecian robes, and they chat contentedly among themselves and with a couple of passers-by who were absent from the original. The picture has been heavily influenced by the conventions of European art – the rocks to the right could be by Salvator Rosa, while the trees convert the barren countryside into a pastoral landscape by Claude Lorrain.

Nevertheless, Hawkesworth's critics accused him of presenting Tahiti as a tainted and immoral paradise. His spicy descriptions of Tahitian erotic rituals were, they said, unsuitable for British readers. The preacher John Wesley was horrified by accounts of the *Endeavour* voyage, although he seems to have been appalled not so much at the displays of sexual abandon, but more that they were performed by light-skinned people. '"Men and women coupling together in the face of the sun, and in the sight of scores of people! Men, whose skin, cheeks, and lips are white as milk"', he reported reading; 'Hume or Voltaire might believe this, but I cannot.'[71] The satirical poets revelled in their images of innocent young women being corrupted. This is a typical taster:

One page of Hawkesworth, in the cool retreat,
Fires the bright maid with more than mortal heat
... [and so on][72]

As well as claiming new territories and bringing back new biological specimens, overseas expeditions discovered new societies that challenged British tradition. In Erasmus Darwin's *Loves of the Plants*, Venus smiles down over Tahiti, a reference to the astronomical transit of the planet as well as to the goddess of love. This is the home, reports Darwin, of the plant *Adonis*, named after the handsome Greek god who was the product of incestuous sex and also the object of Venus's passionate desires. Bearing a hundred male stamens and a hundred female pistils within a single flower, *Adonis* epitomised European dreams of free and easy Tahitian love, and so seemed to mock English convictions that monogamous marriages were the only possible source of social stability and happiness.

* * *

Accounts of the *Endeavour*'s voyage convinced patriotic British Christians that it was their duty to rescue Tahitians from their inferior conditions. Missionaries – some of them helped by Banks – travelled out to convert and educate people in their home environment. Other reformers preferred the converse technique – bringing islanders back to Britain for an intensive course of Western civilisation. Banks's protégés Tupaia and Tayeto had succumbed to foreign diseases, but as a result of Cook's second voyage to the Pacific, Omai, a young man of about twenty, was brought safely to Britain. Delighted at this latest acquisition, Banks and Solander immediately went down to meet him at Portsmouth. After an embarrassing hitch when Omai failed to recognise Solander because he had put on so much weight, the two Europeans welcomed Omai in their pidgin Tahitian and started to teach him English. It was probably Banks who later commissioned a large conversation piece (Figure 14) showing him with Omai and Solander in a comfortably furnished room whose rural outlook suggests they were posing in Banks's Lincolnshire home.

Banks and Sandwich organised Omai's travel schedule and publicity arrangements. Sometimes he must have felt like the human equivalent of a performing seal. After only a few days of intensive

Figure 14. 'Omai, Banks and Solander'. Oil painting by William Parry (1775–6). (National Portrait Gallery.)

rehearsals, he was dressed up in a brown velvet coat and white satin breeches and presented to George III. Omai had been warned by his Pacific friends that the Europeans planned to kill and eat him, so he was probably alarmed to be given a smallpox inoculation taken from a woman with 'several large pustules on her face'.[73] Inoculation was still a risky procedure, and Omai's new British friends prepared him for the illness that followed. Several weeks later, after Omai had recovered, Banks took him on tour round London's aristocratic mansions and metropolitan dining tables.

Omai became the darling of élite society, fêted by celebrities such as Johnson and the Duchess of Gloucester. He was escorted to Britain's greatest spectacles – theatres, the House of Lords, the University of Cambridge – although it is unclear who was meant to be being entertained during these excursions. The novelist Fanny Burney marvelled at Omai's fine clothes and sword, and in her enthusiasm overcompensated by insisting that 'He makes *remarkable* good bows – not for *him*, but for *anybody* ... He eat heartily and committed not the slightest blunder at table.' Overriding his protests, she forced him to sing a Tahitian song and then derided his 'savage' music. Dinner party guests amused themselves by laughing indulgently at Omai's fascination with everyday objects such as magnifying glasses and ice, and when he won at chess or backgammon 'admired at the savage's good breeding'. His accent provided a constant source of hilarity, and historians still repeat the cheap jokes made at his expense.[74]

Banks took over the responsibility for this Pacific student of British culture, at first acting as host himself and later arranging for the government to install Omai in London lodgings. As Omai visited different country estates with Banks, he was forced to participate in impromptu botanical expeditions. A twelve-year-old boy later recalled trips when 'we never saw a tree with an unusual branch, or a strange weed, or anything singular in the vegetable world, but a halt was immediately order'd: – out jump'd Sir Joseph ... and out jump'd Omai after us all'. Nevertheless, Omai does not seem to have been over-enthusiastic. 'Sir Joseph', the retrospective account continues, 'explain'd to us the rudiments of the Linnaean system in a series of nightly lectures, which were very short, clear, and familiar; – the first of which he illustrated by cutting up a cauliflower, whereby he entertain'd the adults (Omai excepted) as much as he delighted the younkers ... I never see a boil'd cauliflower without recollecting the raw specimen, and the dissecting knife, in the hands of Sir Joseph; and thinking on fructification, sexual system, pericarpium, Calyx, corolla, petals &c. &c. &c.'[75]

Although Omai was schooled in polite conversation and behaviour, he was not taught to read fluently. Perhaps, as one critic commented, Banks wanted 'to keep him as an object of curiosity, to observe the workings of an untutored, unenlightened mind'.[76] In Figure 14, one of several portraits of Omai, he is shown being inspected by Banks and Solander, who worked at the British Museum. This portrait presents Omai both as a noble savage and as a strange sample to be scrutinised, classified and catalogued. His flowing white robes, erect posture and bare feet recall a patrician Roman, yet his dark skin and the tattoo on his hand mark him out as a primitive creature. He gazes almost arrogantly out of the canvas, the plain cloth sculpted into sumptuous folds.

Compared with the portly Europeans, Omai appears vigorous and muscular, yet he has been relegated to a lower if exotic order by feminising him. His clothes resemble a woman's dress, and the ostentatious decorations on his hand suggest an attention to appearance more typical of women or effete Macaronis than sober men of science. The *Endeavour* travellers had been fascinated by the tattoos decorating the semi-naked bodies of the local people they watched. In England, Omai's intricate swirling tattoos were usually concealed beneath his clothes, but after a swimming expedition, his companion remarked that 'the tawny Priest ... look'd like a specimen of pale, moving mahogany, highly varnish'd; not only varnish'd, indeed, but curiously veneer'd'.[77]

A mahogany 'specimen' – that is also how Banks and Solander seem to be regarding Omai in this picture (Figure 14), as he stands in silence to be examined. Dressed in a flamboyant red jacket, the overweight Solander sits at his desk while Banks, an intermediary wearing sober colours, points out Omai's tattoo to Solander as well as to us, the viewers. As a good Linnaean, Solander recorded Omai's physical data to accompany the descriptions of all the plants, animals and other natural curiosities collected together at the British Museum. 'He is very brown, allmost as brown as a Mulatto', Solander wrote to a friend. 'Not at all hansom, but well made. His nose is a little broadish …'

Since Omai was at the most semi-literate, we have no first-hand account of his experiences either in England or Tahiti. According to Solander, he had agreed to visit England because at home people laughed at his 'flatish Nose and dark hue, but he hopes when he returns and has many fine things to talk of, that he shall be much respected'.[78] Although many British people were fascinated by Omai, some critics objected to his being decked out like a Macaroni and paraded around the country. In addition, philanthropists protested that he should be given a Christian education. Judging from the frequent comments on his courtesy, it seems that

Omai rapidly learnt to conceal his opinions and emotions beneath a veneer of elaborate politeness. Whatever his feelings, this experiment in cultural indoctrination ended after a couple of years, when Omai was sent back home with Cook's third expedition to the Pacific.

To show how much his European friends cared for him, Omai returned laden down with extravagant gifts. Sandwich ordered up a custom-made suit of armour from the Tower of London, and Banks provided the clothes, cutlery and furniture to which he had become accustomed during his stay. Two drums were provided for improving Tahitian music, while other presents were designed to help Omai stun his friends with the marvels of Western technology – fireworks, miniature horse-driven coaches, a hand-organ, an electrical machine.

Omai's close relatives were overwhelmed with joy on his return – presumably they had given up hope of ever retrieving him from the foreign invaders. Disregarding Cook's paternalistic advice, Omai distributed his strange acquisitions among his friends, and tried to manipulate the English travellers into helping him liberate the island from some neighbouring invaders. However, the Europeans interpreted events differently. From their vantage point, the behaviour of Omai and his compatriots lived down to expectations. According to the navigators' accounts, when Cook's ship landed the

Tahitians showed little interest in Omai until he produced some red feathers he had acquired during an earlier stop at some other Pacific islands. Instead of nails, red feathers now apparently became the currency for the sailors to buy local women on Saturday nights, when Cook allowed them 'to drink to their feemale friends in England, lest amongst the pretty girls of Otaheite they should be wholy forgoten'.[79] With little knowledge of local etiquette, the visitors inferred that Omai had bought allegiance by handing over his possessions to the grasping Tahitians. For the last ten years, Cook sadly concluded, Europeans had been trying to help the islanders better themselves, but in vain – they would insist on retaining their own customs.

Even after Omai had been despatched to the other side of the world, hopefully spreading the virtues of British civilisation, he still provided valuable entertainment material in England. Cook was killed in Hawaii, but one of his naval officers published an anonymous account of Omai's glorious arrival in Tahiti, describing how he paraded round the harbour on a horse, armed with his military presents as though he were St George setting out to kill the dragon. To accompany this lurid text, an imaginary picture showed Omai firing a gun over the heads of spectators as they fled in horror. He had, it seems, truly learnt how to behave like a British gentleman.

Inspired by this extraordinary tale, in 1785 the Theatre Royal decided that its Christmas panto-mime would be *Omai: Or, a Trip Round the World*. As far as the box-office was concerned, this was a brilliant choice, since the extravagant performance was a resounding success. The plot revolves around Omai who, in the fairy-tale version of his life on stage, wins the heart of Londina during his visit to England and triumphantly takes her back to Tahiti. With its elaborate costumes and scenery, this farce reinforced the audience's conviction that King George's Island was an exotic sexual paradise. All the Pacific islands should, a mad prophet declares towards the end of the play, pay tribute to King Omai because he is 'the owner of fifty red feathers, four hundred fat hogs, and the commander of a thousand fighting men and twenty strong-handed women to thump him to sleep'.[80]

· CHAPTER 6 ·

IMPERIALISM AND INSTITUTIONS

It is impossible to conceive that such a body of land [Australia], as large as all Europe, does not produce vast rivers, capable of being navigated into the heart of the interiors; or, if properly investigated, that such a country, situate in a most fruitful climate, should not produce some native raw material of importance to a manufacturing country as England is.

Letter from Joseph Banks to John King, 15 May 1798

If Banks had had his way, Iceland would have joined the British Empire. He spent about six weeks there in 1772, soon after his return from the Pacific, repeatedly complaining about the cold weather and his hosts' lack of humour. Returning laden with minerals, notebooks and Icelandic manuscripts, he promptly printed a business card showing his name – Mr Banks – above an outline of the island, and donated some lava that had been used as ship's ballast to Kew Gardens, where it formed the base for a much-admired mossy carpet. For the rest of his life, Banks kept up the political contacts he had made during this supposedly scientific expedition.

Thirty years later, Banks had levered himself into a far more powerful position. Now he was able to intervene in military tactics during the Napoleonic wars, and he recommended that Britain should seize Iceland from Denmark. As well as making all the obvious arguments about the advantages of acquiring cod fisheries and naval bases, Banks suggested that Iceland was inherently British. 'No one who looks upon the map of Europe can doubt that Iceland is by nature a part of the group of islands called by the ancients "Britannia"', he insisted; 'it ought to be a Part of the British Empire, which consists of every thing in Europe accessible only by seas.'[81] Although Banks never did manage to claim this Danish territory for Britain, he exerted enough influence for an order to be passed banning the British Army from attacking Icelandic ships.

Throughout his long reign as President of the Royal Society, Banks strengthened the ties between science, trade and the state. A skilled diplomat, he excelled at pointing out to wealthy institutions the advantages of patronising science. For instance, in 1801 he persuaded the East India Company to provide £1,200 for a Pacific mapping expedition, instructing the explorers 'to Encourage the men of Science to discover such things as will be useful to the Commerce of India & to find new passages'.[82]

Similarly, by giving advice to the government, Banks hoped to prise funds out of reluctant officials. The Royal Society was unsubsidised, but – especially after the French Revolution – Paris's Royal Academy was supported by state funding. In a pleading letter sent to the British government, Banks played on national pride by stressing how generously foreign countries were supporting scientific research: 'the Academies of the Nations who are our Rivals in Sciences are cherished by their respective Governments at an expence which to their Lordships ... may appear incredible. The Royal Academy of Sciences at Paris have Elegant & spacious apartments ... & Berlin & Petersburg &c the Flourishing Academies are in like manner maintained at Considerable expence.'[83]

In France, Britain's traditional enemy, politicians and scientific researchers were linked together by a strong and formal bureaucracy. In comparison, Britain still operated through the old school tie network. Banks used his extensive contacts to worm his way into the circles of power where decisions were made. One particularly important political patron was the Earl of Sandwich, head of the Admiralty – they had been neighbours in London, both enjoyed fishing, and they even shared the same mistress for a time. It was Sandwich who had engineered permission for Banks's journeys to Newfoundland and Australia, and he later turned to

Banks for advice about organising other expeditions. Sandwich did much to convince the British establishment that scientific and imperial exploration could be profitably linked. For example, he passed on to King George III Banks's recommendation that New Zealand flax should be cultivated in Britain, and pointed out that an Arctic voyage being planned by the Royal Society might advantageously open up new British trade routes.

Constantine Phipps, Banks's friend from Eton and his companion to Newfoundland, also pulled strings for science during his long political career. In the early stages of their mutual rise to eminence, Banks campaigned for Phipps by exploiting his friendship with Sandwich and his prestige as a Lincolnshire landowner. Later on, Phipps reciprocated by backing several of Banks's ambitious projects, such as his plans to explore the American coast and to revolutionise the Caribbean economy by importing breadfruit from Tahiti.

Banks's wealth lay in his country estates, and in political affairs he instinctively inclined towards the conservative side. On the other hand, he often insisted that he was not himself interested in party politics – although he had no qualms about benefiting from his political friendships. Stressing his independence in this way helped Banks to strengthen his relationship with George III, which had started soon after his return from the *Endeavour* voyage. As one of his allies remarked, 'Sir Joseph's political principles, too, those of a high tory, were much to the Monarch's liking; and a country gentleman who never troubled himself with Parliamentary life, nor ever desired to rise above the rank he was born to, was sure to find a friend in His Majesty.'[84] Banks's critics were cynical about this carefully cultivated royal friendship. In Figure 7, Gillray drew a crown in the sun to emphasise how the Great South Sea Caterpillar basked in the warmth of royal favour.

George III had summoned Banks to Windsor immediately after his arrival back in England, and soon put him unofficially in charge of Kew Gardens. Only five years older than Banks, the King welcomed his advice but also valued his friendship. Although Banks was always the subordinate, for about 30 years they were as intimate as a royal patron and his protégé can be. In 1787, for instance, George III commiserated with Banks on feeling ill: 'The King is sorry to find Sir Joseph is still confined; and though it is the common mode to congratulate persons on the first fit of the Gout, he cannot join in so cruel an etiquette.'[85] Only a couple of years later, it was the King's turn to be the patient, when, suffering from an undiagnosed and possibly hereditary illness, George III endured his first bout of insanity. As he started to recover, he summoned Banks to take daily walks with him through the Royal Gardens at Kew.

An ideal opportunity for Banks to boast about the benefits of botany – and after the King was well again, Banks took full advantage of the royal fascination with plants that he had nurtured.

This close relationship was further consolidated by their shared interest in agriculture. Just as Banks relied on the income from his lands, so too, the major source of British wealth before the industrial revolution lay in its farms. In addition to his roles as an imperial explorer (Figures 6 and 9) and scientific administrator (Figures 8 and 14), Banks was also a wealthy landowner. The Corporation of Boston (in Lincolnshire) commissioned Phillips for another version of his portrait as President of the Royal Society (Figure 8). Although Phillips painted Banks in the same pose, this time he showed him wearing the county's military uniform and holding a fen drainage scheme in his hand.

By retrieving the marshes for pasture, Banks benefited Lincolnshire farming but also increased his own wealth. Improving the land also meant improving his own position. Sheep were vital for converting this reclaimed territory into fertile fields that would yield large crops of wheat. In the oil painting of Figure 15, Banks (fourth from the door frame on the left) is shown participating in the sheep-farming activities of central England. Like his fellow members of the landed gentry, Banks sports a dignified hat and his tailored jacket strains round

Figure 15. 'Ram letting from Robert Bakewell's breed at Dishley, near Loughborough, Leicestershire' (1810), by Thomas Weaver. (Tate Gallery.)

his well-fed bulk. Although George III governed a larger territory than Banks, both men felt a paternalistic responsibility for the people they ruled, and were committed to making agriculture more profitable. As property-owning conservatives, they both felt that it was in the local as well as in the national interest to cut down on expensive wool imports in order to encourage English production. In 1781, at the start of a project to improve the quality of British wool by smuggling in Spanish sheep, the King knew who he wanted. 'Sir Joseph Banks is just the Man', he informed his staff; 'Tell him from Me that I thank Him, & that his assistance will be most welcome.'[86]

As head of the Royal Society and confidant of the King, Banks was in a unique position to show how scientific research could make Britain's growing empire even more profitable. All the long walks round Kew Gardens placating an unbalanced sovereign paid off. By acting as an intermediary with the court, Banks levered himself into a secure position as an essential scientific adviser to the government. Banks took advantage of his friendship with George III to forge tighter bonds between science, the state and Britain's trading empire. Thus he made himself an expert on tea-growing because he wanted to cut down the expense of Britain's imports from China. Writing long letters of technical advice to the East India Company, he encouraged them to grow tea on British land in India, and also persuaded the King that a plant-gathering expedition to China would bring 'Reel advantage to this Country & her Colonies, as well as much improvement to the Science of Botany & to the Botanic Gardens at Kew, which are now a favourite Object of recreation to the whole of the Royal Family.'[87]

* * *

While Banks was still at Oxford, his mother lived in

Chelsea, then a fashionable suburb with open fields. Amidst all his other metropolitan entertainments, Banks enjoyed visiting the Chelsea Physic Garden, which was run for 48 years by Linnaeus's friend Philip Miller. By importing plants from all over the world, Miller had quintupled the collection. When Banks strolled round its beds, the Garden had expanded its role – no longer just a source of medicinal plants, it had also become an international centre for botanic research. After Miller died, Banks bought his herbarium (collection of dried plants), which was apparently so large that it took a fortnight to move.

Miller's garden was small, and his primary aim was to make it useful. In contrast, although the superintendent at Kew was one of Miller's trainees, the Royal Gardens were far larger, better funded, and had originally been designed for pleasure. During his long regime as the King's adviser, Banks converted Kew into the world's leading botanic garden, making it a central clearing house for an imperial trade in agricultural development. Located at the hub of the British Empire, Kew contained plants from all over the world, and many of them were grown for their commercial potential as well as for their scientific value. There was a three-way traffic in plants. Using his extensive correspondence network, Banks scoured the world for useful crops to cultivate in Britain; simultaneously, he altered the patterns of international vegetation by exporting plants to British colonies, and also by moving them round the empire from one country to another.

Apart from exceptional curiosities like Omai, during the 18th century most settlers in Britain's developing empire travelled outwards from the centre. In contrast, plants were being carried back in the opposite direction. Under Banks's care, Kew Gardens expanded rapidly, and by 1788, 50,000 trees and plants were growing in the beds and hothouses. As well as all the fuchsias, magnolias and other exotics, some individual plants became world famous – a delicate Venus flytrap from South Carolina flourished at Kew even though the one owned by Buffon in Paris had withered away, and an exceptionally striking flower was diplomatically named *Strelitzia regina* after the queen. As Banks transformed this pocket of English countryside into a foreign paradise, he boasted that 'our King at Kew & the Emperor of China at Jehol solace themselves under the shade of many of the same trees & admire the elegance of many of the same flowers in their respective gardens'.[88]

But Banks was primarily interested in economic botany, and his early transplants to Kew included two potentially profitable plants from New Zealand – flax and spinach. By emphasising the commercial value of botany, Banks persuaded George III to pay professional collectors. In addition, he received contributions from an international network of unofficial botanists, who included politicians as well as soldiers and seamen, merchants and missionaries. To solicit still more gifts, Banks thoughtfully named some plants after their donors: an Ethiopian plant is still called *Brucea* after James Bruce, the Fellow of the Royal Society who traced the Blue Nile to its source. Banks was determined that Kew should boast a more impressive collection than any other country – especially France. Hearing of a forthcoming French expedition to Australia, he immediately sent out a British collector to gain 'an opportunity of collecting plants, which could by no other means be obtained; & of enriching the Royal Gardens at Kew with plants which otherwise would have been added to the Royal Gardens at Paris'.[89]

Under Banks's supervision, collectors brought back thousands of exotic bulbs, seeds and plants – this was the period when monkey puzzle trees and evergreen sequoias first came to Britain. There were, however, numerous disasters. One recruit sent back hundreds of specimens from Africa but succumbed to the climate in Canada; another was – much like Banks on the *Endeavour* – appointed as the naturalist on a naval ship, but all the plants he had carefully taken on board died from lack of water while he was imprisoned after a fight with the captain. Banks liked to keep a tight control over his delegates. When one collector threatened to settle in Australia, Banks stormed: 'I did not take him to beget a family in New South Wales. I fear if he is not more active than is compatible with a married life I must get rid of him.'[90]

Banks also arranged for plants from one country to be tried out in others with similar climates. The Board of Agriculture came to recognise that Banks was the man who could advise them on questions of imperial botany – would seeds from Sumatra grow in the Caribbean, for instance, or how might sugar production be improved in Surinam? Moving crops from one part of the world to another could dramatically increase their value.

One of Banks's most ambitious projects was to transplant breadfruit from the southern Pacific to the West Indies. This plan was especially popular among plantation owners, who hoped that bread-fruit would provide a cheap way of feeding their black slaves. After Banks had persuaded the Admiralty and the Home Secretary that his idea would work, he was allocated a Royal Navy ship – the *Bounty*, to be commanded by William Bligh. Converting the *Bounty* into a floating garden, Banks made it clear that the survival of his Tahitian plants was far more important than the comfort of the naval officers: the trees even had first call on the fresh water needed to wash off the salt from the damp air. Presumably remembering his own experiences on the *Endeavour*, Banks ordered up poison for the rats and cockroaches, insisting that 'the crew must not complain if some of them who may die in the ceiling make an unpleasant smell'.[91]

Bligh's expedition to Tahiti was a total failure. Even under the most difficult circumstances, Cook had managed to keep order on his ships. But Bligh lacked Cook's diplomacy, and even before the *Bounty* reached Tenerife, he was hardly on speaking terms with his men. Nevertheless – at least, according to Bligh's version of events – on Tahiti he did coax the crew into carrying out the heavy labour needed to load hundreds of breadfruit trees, and together they set off for the Caribbean. But they never arrived. This was the voyage of the famous mutiny, when the sailors took over the *Bounty*, abandoning Bligh to find his own way home without his botanical cargo.

Surprisingly, a few years later, Banks and the Admiralty trusted Bligh enough to send him back to the Pacific again. This time the botanical aspect of the trip was a great success. Over 2,000 Tahitian breadfruit trees were planted in sawn-down casks, and many of them survived their journey to flourish in their new colonial home. Economically, however, the scheme was initially less profitable – the West Indians were reluctant to eat this foreign food that English landowners were trying to impose on them, and children played football with the dried-out surplus fruits. But Tahitian breadfruit is now known as a Caribbean staple, exported as a regional speciality for sale in London's street markets.

Banks superintended an international network of botanic gardens that made this redistribution of the world's crops possible and also extended Britain's power. Declaring that Kew should become 'a great botanical exchange house for the empire', Banks converted the Royal Gardens into the head office of an international agricultural chain committed to commercial development. For instance, with his help, George III resuscitated the garden in St Vincent to act as a temporary storehouse both for American plants being sent to Kew and for Asian and Pacific plants being imported to the West Indies. As Banks made clear in his instructions to the superintendent, colonial botanic gardens were important for Britain's economy. They would prove their worth, he promised, by allowing 'the introduction of many articles of value in a commercial or medicinal view, only produced in foreign settlements, & not to be procured by the British, but at very high prices'.[92]

By the early 19th century, gardens had become a standard symbol of colonial conquest. As part of his schemes to make tea cheaper for British consumers by growing it in India, Banks became intimately involved in proposals to establish a Botanic Garden in Calcutta, and later arranged for it to receive samples of Australian flax. He also promised the War Office that by helping to provide food, this Garden would make the Indians 'wonder how their ancestors were able to exist without them & revere the names of their British conquerors to whom they will be indebted for the Abolition of Famine'.[93]

Ceylon (now Sri Lanka) is another good example of Banksian imperial botany. In 1810, several years after seizing Ceylon from the Dutch, the British rulers rescinded a law banning European farming methods. Banks sent out a gardener from Kew with a double mission – to work 'for the benefit of the commercial interests of the island, and for the advancement of the Science of Botany'. By creating a botanical garden, a miniature Kew under the King's patronage, the island's developers demonstrated that foreign crops such as coffee would flourish and bring money into the colony. As the local economy boomed, British imperialists boasted that Ceylon was a marvellous illustration of their enlightened rule.[94]

Botanic gardens also provided a way for British immigrants to garner local expertise. Apart from his stay in Tahiti, Banks had not had enough close contact with indigenous peoples to learn much about their own knowledge. As more permanent settlements were established throughout the world, he encouraged residents to export foreign skills back to Britain. He recommended using the Ceylonese Garden to study herbs prescribed by local doctors so that British medicines could be made more effective. In China, Banks resorted to industrial espionage for tying together science and the state – he campaigned to reinstate the British embassy in Peking because he needed a cover for craftsmen who would inform him about China's methods of producing teas and porcelain. The appreciative ambassador collaborated, sending back porcelain samples 'with the view of having them compared under the eyes of Chemists and skilful artists with the materials used in England'.[95]

Banks also sent European plants to be cultivated abroad. With the cooperation of the Home Office, he transplanted plants and animals to places in the opposite hemisphere with a similar climate. For example, he shipped Mediterranean crops to be grown in New South Wales, and Cook took pigs to New Zealand (where they ran wild and are still called Cookers). As a consequence, regions of the world that lie far away across the oceans started to resemble Europe.

In 1776, the influential Scottish economist Adam Smith suggested that: 'The colony of a civilised nation which takes possession, either of waste country, or of one so thinly inhabited, that the natives easily give place to the new settlers, advances more rapidly to wealth and greatness than any other human society.' Banks was instrumental in introducing Smith's 'new settlers' to Australia, New Zealand and other British colonies. His immigrants included crops and animals as well as people. To satisfy imperial requirements, British invaders – the grains and meat-animals most in demand by European consumers – displaced the original inhabitants. Distant countries became neo-Europes where sheep and cows grazed on the hillsides and farmers cultivated wheat, barley, rye and potatoes – imports regarded as foreign exotics by the local people. (Banks and his successors also transported less useful goods: in the southern hemisphere, dandelions and house cats forced kangaroo grass and kiwis into retreat, while tuberculosis, smallpox and sexually transmitted diseases savagely reduced human populations.)[96]

Financial gain was a major objective. Arthur Young, one of Banks's allies and Britain's leading agricultural expert, preached that 'the best use the land can be put to, is to cultivate THAT crop, whatever it be, which produces the greatest profit VALUED IN MONEY'. Like other wealthy landowners, Banks made the farming techniques on his Lincolnshire estates more efficient to help feed the country but also to maximise his own income. Similarly, he suggested that by growing the crops most needed in Europe, the colonies as well as Britain would gain financially. India, he argued, should stop exporting expensive material and provide Britain with the raw cotton it needed for its own factories to make a profit. This trade would, he insisted to the East India Company, benefit both partners and bind the empire together: 'A colony such as this, blessed with the advantages of Soil, Climate, Population so eminently above its Mother Country, seems by nature intended for the purpose of supplying her fabrics with raw materials; & it must be allowed that a Colony yielding that kind of tribute binds itself to the "Mother Country" by the strongest and most indissoluble of human ties, that of common interest & mutual advantage.'[97]

* * *

Botany was Banks's prime passion, but he became involved in many other projects of imperial science as well. Through his international network of contacts, he developed all sorts of schemes that would – to use a favourite Enlightenment parallel – *improve* agricultural and industrial production and *improve* human beings. From his base at Soho Square, Banks controlled British institutions like Kew Gardens and the Royal Society, and also exerted a strong influence on their colonial offshoots, the botanical gardens and scientific societies being established throughout the growing British Empire. As well as the Admiralty and the government, private organisations like the East India Company and the Sierra Leone Company also turned to Banks for advice. Through diplomatically exchanging expertise for financial patronage, Banks helped to ensure that science, trade and commercial expansion became indissolubly linked together.

Another way that Banks consolidated his power was by joining committees designed to sponsor imperial exploration. In 1788 he was among the founding members of the Association for Promoting the Discovery of the Interior Parts of Africa. As usual, motives were mixed. Among its idealistic goals, the Association declared it would alleviate ignorance (European as well as African). Since the original committee was dominated by Fellows of the Royal Society, scientific research was high on the agenda, but anti-slavery campaigners were also aiming to combat the slave trade. Yet, inevitably, exploration was intimately intertwined with commercial interests. As well as looking for sources of raw materials, Britain wanted to create markets for the industrial goods it was producing. And since the French were also colonising Africa, there was a strong element of competitiveness.

The preliminary meeting was held at Banks's London home, and he energetically pulled in government support for his new Association. Within a few years, aware that French activities in Africa were increasing, Banks's aims had become explicitly territorial. In order, he told the government, to export manufactured products, mine gold and benefit the local people, Britain should convert them to Christianity and 'secure to the British Throne, either by conquest or by treaty, the whole of the coast of Africa from Auguin to Sierra Leone, or at least to procure the cession of the River Senegal, as that River will always afford an easy passage to any Rival Nation'.[98]

Like many of his colleagues, Banks thought of himself as an enlightened man who was improving rather than exploiting Britain's colonial possessions. However, local inhabitants did not always appreciate his Association's involvement, which they saw as interference. 'They express on all occasions', ran a report to the African Association, 'a conviction that the soil and the country is their own, saying this is not white man's country, this belong to black man, who will not suffer white man to be master here ... They have no intention of embracing Christianity, saying they are too old for that.' Convinced that their superior influence could only bring benefits, European imperialists felt justified in quelling resistance to dominate other people for their mutual advantage. As George III advised his Prime Minister, barbarians could not be governed 'with the same moderation that is suitable to a European civilized Nation'.[99]

Other beneficiaries of Banks's schemes also

objected to the way that they were being treated. Lincolnshire farmers rioted after they lost their land through the changes he had introduced, and in London the army had to protect Banks's house from angry crowds protesting about the manipulation of grain prices. As a privileged country gentleman, Banks adopted much the same reforming zeal for upgrading his own property and its inhabitants as he did for converting the rest of the world to British customs. At the same time as George III was backing Banks in his plans to improve the world, these two agricultural landowners were also collaborating on a scheme to develop England's wool industry. Both of them stood to gain by increasing wool production, which would – hopefully – make Britain more self-sufficient and also revive the flagging profits from Banks's estate.

Banks belonged to a group of landowners in central England whose wealth depended on breeding sheep for wool and meat. Meetings like the ram-letting shown in Figure 15 enabled farmers to cross-breed sheep and establish new strains, but they were also designed to advertise an owner's wealth and to provide a forum for discussing agricultural politics. Here these portly men in their expensive sombre clothes are clearly distinguished from their subordinates who are carrying out the physical labour – the stockmen (wearing suspiciously clean white smocks) who are handling the pinkish sheep. As they survey their workers, these wealthy farmers might well be debating a long-standing contentious question: should the government control the prices of agricultural products? Although he claimed to be uninterested in politics, Banks became one of the leading lobbyists for the landed gentry who – like the sheep-breeders in this picture – wanted to protect their own income by exporting raw wool and keeping out cheap foreign imports. This was a major political issue, since at the end of the 18th century, woollen goods accounted for about a quarter of the country's exports, more than iron and cotton combined. In line with Banks and his agricultural community, George III also wanted to boost home production and avoid having to buy materials in from abroad.

Lincolnshire weavers had traditionally worked with the local heavy wool, but many customers preferred the finer wool produced by Spanish merino sheep. One obvious solution was to rear merinos in Britain, but critics objected that the colder climate would make their wool become coarse. Banks disagreed. First he persuaded a French colleague (suitably rewarded with imperial produce – English turnips, Chinese hemp seed and an Australian kangaroo) to send him a pair of pure merinos, and then crossbred these with sheep gathered from the farms of his landowning friends.

In particular, Robert Bakewell was a pioneer in agricultural experimentation, and Figure 15 shows Banks attending a ram-letting at his farm, where the gentlemen are bidding for a season's use of Bakewell's expensive rams. Famous for his new Leicester breed, Bakewell claimed to have doubled England's mutton production (although here his sheep are probably painted rather larger than life), and he was also trying to develop fine wools.

Kew was the world's leading botanical centre, but there was also ample grazing ground for a sheep-breeding trial. Conniving with George III, Banks concocted several schemes to get hold of some merinos without the knowledge of Spanish farmers, who would resent losing their export trade. After some botched attempts, he successfully used a network of merchants in Portugal to smuggle some sheep into England. After walking for a couple of weeks from Dover, most of them arrived safely at Kew and Windsor and were later joined by further consignments.

Apart from some initial hiccups, the flock flourished and – just as Banks had insisted – kept their fine wool despite the English climate. As the King strolled round the grounds with Banks, the finer points of sheep breeding and wool production formed an important topic of conversation. George III remained deeply involved in this joint project, and by giving away over 200 sheep to breeders around the country, could feel that he was bene-fiting the British wool industry while ostensibly remaining ignorant of Banks's shady import manoeuvres.

In 1804, Banks arranged to sell some of the royal merinos by auction. Eleven of them were illegally exported to New South Wales, where they helped to found Australia's massive sheep industry. At first, Banks got it completely wrong: he predicted that this venture was bound to fail. Nevertheless, he soon started advising the government how to parcel up the land for private farming schemes. Because he was one of the very few British people who had ever been to Australia, he was often consulted and, even though he had no official post, came to wield enor-mous political influence over the continent's devel-opment. As he entrenched himself within Britain's élite, Banks confidently assured a newly appointed colonial governor that he would protect Australia while 'Ministers, fully occupied with the business of carrying on a calamitous war, have ... much neglected the Interests of your Establishment, my favorite Colony'.[100]

Sheep were not the only Australian settlers organised by Banks. In 1779, a House of Commons committee had turned to the 34-year-old expert for advice on where to send British prisoners. Australia, reported Banks, was the perfect place. Unlike the New Zealand Maori, the indigenous inhabitants were terrified of Europeans and would offer no resistance. Furthermore, he optimistically claimed, the land was so fertile that within a year the convicts would be able to support themselves. While the government vacillated, Banks continued with his behind-the-scenes activities, boasting that the new colony would soon be exporting indigenous flax for sails as well as cultivating tea, silk and spices transplanted from other imperial outposts. Seven years later, Britain's overflowing prisons were approaching crisis point, and – despite its many strategic disadvantages – Botany Bay became by default the destination of the first convict ships. The potter Josiah Wedgwood celebrated the event by producing a medal with clay that Banks had sent him from Sydney Cove.

Banks made sure that he was intimately involved in the colonisation of New South Wales. He acted as go-between for ship-owners trying to get contracts for sending out further consignments of convicts, and encouraged research into the rewards of mining for coal and other minerals. Because he hoped that Australia would benefit Britain, he instructed his collectors to find 'objects both in the vegetable & mineral kingdoms hitherto undiscovered, that will, when brought forward, become objects of national importance, & lay the foundation of a trade beneficial to the mother country with that hitherto unproductive colony'.[101]

In addition, Banks made himself responsible for transforming the colony into a distant version of Europe. He supervised the plants that were transported with the First Fleet of prisoners, and subsequently kept sending out crop specimens which permanently altered Australian vegetation – wheat, for instance, and vegetables from the south of France. As one of his plant-collecting protégés remarked, 'Sir Joseph was considered the Father and Founder of the Australian Colonies'.[102]

· CHAPTER 7 ·

HEROES AND HEMISPHERES

*What Great Britain calls the Far East is to us
the near north.*

> Robert Gordon Menzies,
> Sydney Morning Herald, 1939

Confined to his study with gout, and aware that
he was losing his grip over the Royal Society, in
his old age Joseph Banks must have reflected on
his future fate. How would he be celebrated, he
wondered – as a great botanist, a pioneer explorer
or a powerful scientific administrator? Or did he
hope to be remembered for his fine Lincolnshire
wool and his Australian solution for Britain's over-
crowded prisons? In his gloomier moments, Banks
probably anticipated his actual fate during the 19th
century: he was dismissed as an aristocratic autocrat
by the young chemists, physicists and mathematicians who wanted to revolutionise British science.

Banks had, after all, witnessed the posthumous
decline of his own hero, Carl Linnaeus, the man
whose travels in the Arctic north had inspired
Banks to sail with the *Endeavour*. After his return to

Europe, Banks never did keep his promise to visit the sick and elderly botanist in Uppsala. Once in a position of power, Banks himself helped to displace Linnaeus. His own research counted against the Swedish model of human races, and he encouraged classifiers to develop different systems for cataloguing their specimens. Within only a few years of his death, Linnaeus had become the laughing-stock of educated Swedes. Crates of artefacts shipped back by his students were left unpacked, and even his former followers wrote parodies mocking themselves as a 'Grass-hunter troop under the so very famous Sir and Knight von Linné'.[103]

But although Banks never knew about it, Linnaeus's reputation was gradually resuscitated. As Swedish nationalism grew, he became the country's Romantic icon, the equivalent of England's Shakespeare or Germany's Goethe. Despite having brown eyes and being a notoriously scruffy dresser, Carl Linnaeus (briefly Europeanised as Karl von Linné) acquired blond hair, blue eyes and clothes that matched the colours of the Swedish flag. However, in the 20th century, his reputation faded again as socialist governments discouraged conservative bonds with the past – although a local Linnaean industry does still thrive in Uppsala.

Linnaeus had regarded Lapland as an exotic utopia, the northern equivalent of Banks's Tahiti. During Banks's lifetime, Britons and other southern

Europeans converted Sweden itself into an ethno-graphic holiday destination. Linnaean researchers became tourist attractions who, impoverished, sold off their collections to provide sentimental scien-tific souvenirs. Well into the 20th century, Sami nomads were objects of curiosity, displayed in fairs as primitive savages and symbols of a romanticised alien existence. In the late 20th century, as issues of fourth-world politics developed in importance, Samis aligned themselves with Inuit and other inhabitants of the circumpolar north.

The old political empires have been disman-tled, but scientific imperialism still survives. In the southern hemisphere, the Antarctic zone has – like outer space – been converted into an international science laboratory. In contrast, the circumpolar region of the Arctic Circle has long been divided between the countries lying around its southern edge. Because of the area's high strategic value and rich mineral deposits, powerful nations are reluctant to abandon their claims. Linnaeus was just one of the many Scandinavian, Canadian and Russian explorers who, like Banks, twined together scientific research, commercial opportunism and imperial possession.

Outside Sweden (and biology textbooks), Linnaeus has never been particularly famous. In contrast, Banks is little known in his native country but has become Australia's founding father. The red honeysuckle that Banks collected during his brief visit to Botany Bay was christened *Banksia ericifolia* by Linnaeus's son, and *Banksia* plants were unique to Australia. Banks, botany and Australia are indissolubly tied together, even though he was only there for a few weeks – and even then, often at sea rather than on land.

Yet Banks is a problematic hero for Australia. When news of his death eventually reached the distant colony, a monument was erected in Botany Bay by the governor of Brisbane, an astronomer who had won his appointment through Banks's influence. The plaque tactfully commemorated him as a scientific discoverer who was 'Ardent in the Pursuit of Knowledge', but the local newspaper carried a poem cynically referring to Banks's role in establishing New South Wales as a penal colony that was

> ... *big with virtues (though the flow'ry name*
> *Which Science left it, has become a scorn*
> *And hissing to the Nations), if our Great*
> *Be wise and good.*[104]

It was easy to blame Banks for converting Australia into an overflow prison run with barbaric cruelty. How then could it be possible to admire him as a national icon? Partly because of this ambivalence, Banks's status has fluctuated. Towards the end of the 19th century, when the continent's six separate colonies were fusing together into a single nation, patriotic Australians tried to explain away Banks's disreputable behaviour. He was, they argued, a misplaced star who had merely acted in accordance with the principles of his age. Banks became a symbol of Australia's independence and ability to make original contributions to world science. With the help of government funding, many of his manuscripts went to Australian libraries where they generated a miniature biographical industry.

But towards the end of the 20th century, when Australia was severing its ties with Britain, enthusiasm for Banks waned. Ironically, now it is British scholars who want to rescue him from obscurity. Historians have become disillusioned with old-fashioned stories of great discoverers, and are far more interested in exploring how science has become so central to modern society. Banks provides a marvellous illustration of how science and the British Empire grew rich and powerful together.

Further Reading

Banks, Joseph. *The* Endeavour *Journal* (ed. J.C. Beaglehole) (Sydney: Angus and Robertson, 1962). Banks's original account, with a long introduction.

Carter, Harold. *Sir Joseph Banks 1743–1820* (London: British Museum (Natural History), 1988). The definitive full biography.

Drayton, Richard. *Nature's Government: Science, Imperial Britain, and the 'Improvement of the World'* (New Haven and London: Yale University Press, 2000). Innovative account covering the 16th to the 19th centuries.

Gascoigne, John. *Joseph Banks and the English Enlightenment* and *Science in the Service of Empire* (Cambridge: Cambridge University Press, 1994 and 1998). This two-volume study is the most thorough modern assessment.

Jardine, Nicholas, Secord, James and Spary, Emma (eds). *Cultures of Natural History* (Cambridge: Cambridge University Press, 1996). Excellent survey collection of short articles.

Koerner, Lisbet. *Linnaeus: Nature and Nation*

(Cambridge MA and London: Harvard University Press, 1999). The best biography – scholarly but lively.

O'Brian, Patrick. *Joseph Banks: A Life* (London: Collins Harvill, 1988). Engaging account which focuses on his voyage with Cook.

Schiebinger, Londa. *Nature's Body: Gender in the Making of Modern Science* (Boston: Beacon Press, 1993). Clearly written study of race, gender and science.

Smith, Bernard. *European Vision and the South Pacific* (Oxford: Oxford University Press, 1989). The classic art historical study of European interpretations.

NOTES

1. This introduction is based on R. Porter, 'The Exotic as Erotic: Captain Cook at Tahiti', in G.S. Rousseau and R. Porter (eds), *Exoticism in the Enlightenment* (Manchester and New York: Manchester University Press, 1990), pp. 117–44. (Robertson quoted p. 125, Cook quoted pp. 127, 128.)

2. J. Banks, *The Endeavour Journal* (ed. J.C. Beaglehole), 2 vols (Sydney: Angus and Robertson, 1962), quotations from vol. 1, pp. 276, 289, 281, 282.

3. Quotations from pp. 180, 179, 182 of A. Bewell, '"On the Banks of the South Sea": Botany and Sexual Controversy in the Late Eighteenth Century', in D. Miller and P. Reill (eds), *Visions of Empire: Voyages, Botany, and Representations of Nature* (Cambridge: Cambridge University Press, 1996), pp. 173–93, my major source for this section.

4. Quoted on p. 253 of M. Cohen, 'The Grand Tour: Constructing the English Gentleman in Eighteenth-century France', in *History of Education*, vol. 21 (1992), pp. 241–57.

5. J. Perry, *Mimosa: or, The Sensitive Plant* (London, 1779), pp. iii, vii.

6. Quoted on pp. 22–3 of L. Schiebinger, *Nature's Body: Gender in the Making of Modern Science* (Boston MA: Beacon Press, 1993). My major source for this chapter is L. Koerner, *Linnaeus: Nature and Nation* (Cambridge MA and London: Harvard University Press, 1999).

7. Quoted in Koerner, *Linnaeus*, p. 43.

8. Quoted on p. 61 of P. O'Brian, *Joseph Banks: A Life* (London: Collins Harvill, 1988).

9. Quoted in Koerner, *Linnaeus*, p. 127.

10. Quoted on p. 133 of A. Bewell, '"Jacobin Plants": Botany as Social Theory in the 1790s', *Wordsworth Circle* 20 (1989), pp. 132–9.

11. Quoted on p. 268 of J. Uglow, *The Lunar Men: the Friends who made the Future* (London: Faber and Faber, 2002).

12. J. Lee, *An Introduction to Botany, chiefly extracted from the Works of Linnaeus* (London, 1810), p. xvii (introduction by Robert Thornton).

13. Quotations on pp. 37, 9 of A. Shteir, *Cultivating Women, Cultivating Science: Flora's Daughters and Botany in England 1760 to 1860* (Baltimore and London: Johns Hopkins University Press, 1996), my major source for this section.

14. E. Darwin, *The Botanic Garden* (Yorkshire: Scolar Press, 1973 (facsimile of 1791 edition)), Book 2 (*Loves of the Plants*), pp. 4–5 (Canto I, ll. 51–6). See J. Browne, 'Botany for Gentlemen: Erasmus Darwin and *The Loves of the Plants*', *Isis* 80 (1989), pp. 593–621.

15. *Encyclopaedia Britannica*, quoted on p. 9 of R. Thornton, *The Temple of Flora* (ed. R. King) (London: Weidenfeld and Nicolson, 1981).

16. Quoted on p. 189 of Bewell, '"On the Banks of the South Sea"'.

17. A. Secord, 'Science in the Pub: Artisan Botanists in Early Nineteenth-century Lancashire', *History of Science* 32 (1994), pp. 269–315 (quotation on p. 277).

18. Quotations from p. 26 of H. Carter, 'Sir Joseph Banks: The Man and the Myth', *Bulletin of Local History, East Midland Region* 24–5 (1989–91), pp. 25–32.

19. J. Gascoigne, 'The Scientist as Patron and Patriotic Symbol: The Changing Reputation of Sir Joseph Banks', in M. Shortland and R. Yeo (eds), *Telling Lives in Science: Essays on Scientific Biography* (Cambridge: Cambridge University Press, 1996), pp. 243–65 (quotation on pp. 244–5).

20. Quotations on pp. 48, 55 of O'Brian, *Joseph Banks*.

21. Quoted on p. 9 of J. Gascoigne, *Joseph Banks and the English Enlightenment: Useful Knowledge and Polite Culture* (Cambridge: Cambridge University Press, 1994).

22. M. Shelley, *Frankenstein or The Modern Prometheus: The 1881 Text* (Oxford and New York: Oxford University Press, 1993), p. 7.

23. *European Magazine* 42 (1802), p. 163.

24. Banks, *Endeavour Journal*, vol. 1, p. 113 (Beaglehole's introduction).

25. *European Magazine* 42 (1802), p. 163.

26. Quoted on p. 253 of Gascoigne, *Banks and the English Enlightenment*.

27. Quoted in Bewell, '"On the Banks of the South Sea"', p. 190.

28. Quoted on p. 4 of J. Gascoigne, *Science in the Service of Empire: Joseph Banks, the British State and the Uses of Science in the Age of Revolution* (Cambridge: Cambridge University Press, 1998).

29. J. Wolcot, *The Works of Peter Pindar* (London, 1830), p. 5.

30. Charles Blagden, quoted on p. 85 of J.L. Heilbron, 'A Mathematicians' Mutiny, with Morals', in P. Horwich (ed.), *World Changes: Thomas Kuhn and the Nature of Science* (Cambridge MA: MIT Press, 1993), pp. 81–129.

31. Letters from Banks to Charles Blagden of 7 and 20 December 1819, Royal Society, BLA.b.85–6.

32. Davies Gilbert, quoted on p. 255 of Gascoigne, *Banks and the English Enlightenment*.

33. Letter 110 in N. Chambers (ed.), *The Letters of Sir Joseph Banks: A Selection, 1768–1820* (London: Imperial College Press, 2000).

34. Benjamin Robert Haydon, quoted on p. 7 of D. Shawe-Taylor, *The Georgians: Eighteenth-century Portraiture and Society* (London: Barrie and Jenkins, 1990).

35. G.P. Nuding, 'Britishness and Portraiture', in R. Porter (ed.), *Myths of the English* (Cambridge: Polity Press, 1992), pp. 237–69 (quotation on p. 251).

36. Quoted on p. 212 of J.E. McClellan, *Science Reorganised: Scientific Societies in the Eighteenth Century* (New York: Columbia University Press, 1985).

37. Quotations on pp. 99, 100 of A. Salmond, *Two Worlds: First Meetings between Maori and Europeans 1642–1772* (Auckland: Viking, 1991).

38. Quotation from p. 65 of O'Brian, *Banks* and p. 102 of Salmond, *Two Worlds* (my two major sources for this chapter).

39. Banks, *Endeavour Journal*, vol. 1, p. 168.

40. Quoted on p. 76 of H.B. Carter, *Sir Joseph Banks 1743–1820* (London: British Museum (Natural History), 1988).

41. Banks, *Endeavour Journal*, vol. 1, p. 252.

42. Banks, *Endeavour Journal*, vol. 1, p. 267.

43. Banks, *Endeavour Journal*, vol. 1, p. 269.

44. Banks, *Endeavour Journal*, vol. 1, p. 285.

45. Banks, *Endeavour Journal*, vol. 1, pp. 312–13.

46. Banks, *Endeavour Journal*, vol. 1, p. 386.

47. Chambers, *Letters*, letter 6.

48. Quoted in Salmond, *Two Worlds*, p. 270.

49. Banks, *Endeavour Journal*, vol. 2, p. 51.

50. Banks, *Endeavour Journal*, vol. 2, p. 59.

51. Banks, *Endeavour Journal*, vol. 2, p. 85.

52. Banks, *Endeavour Journal*, vol. 2, pp. 95, 107.

53. Quoted in O'Brian, *Banks*, pp. 169–70.

54. Quoted in Gascoigne, *Banks and the English Enlightenment*, p. 107.

55. J. Boswell, *Tour to the Hebrides with Samuel Johnson*, 1 September 1773 (ed. F.A. Pottle and C.H. Bennett) (London: Heinemann, 1936).

56. Quoted in D. Bindman, *Ape to Apollo: Aesthetics and the Idea of Race in the 18th Century* (London: Reaktion, 2002), pp. 64–5.

57. J. Priestley, *Experiments and Observations on Different Types of Air* (Birmingham, 1790), vol. 1, p. xxxi.

58. Quoted in Gascoigne, *Banks and the English Enlightenment*, p. 136.

59. Quoted on p. 82 of Schiebinger, *Nature's Body*, my major source for this section.

60. Quoted on p. 17 of J.V. Douthwaite, *The Wild Girl Natural Man and the Monster: Dangerous Experiments in the Age of Enlightenment* (Chicago: Chicago University Press, 2002).

61. H. Wallis, 'The Patagonian Giants', in R. Gallagher (ed.), *Byron's Journal of his Circumnavigation 1764–1776* (Cambridge: Hakluyt Society, 1964), pp. 183–223 (quotations on pp. 188, 204.)

62. Banks, *Endeavour Journal*, vol. 1, p. 227.

63. Quoted in Salmond, *Two Worlds*, pp. 112–13.

64. Banks, *Endeavour Journal*, vol. 1, p. 334.

65. Banks, *Endeavour Journal*, vol. 1, p. 351.

66. Banks, *Endeavour Journal*, vol. 1, p. 334.

67. Quoted in Salmond, *Two Worlds*, pp. 87–8.

68. Quoted in O'Brian, *Banks*, p. 91.

69. Quotations on pp. 42, 43 of B. Smith, *European*

Vision and the South Pacific (Oxford: Oxford University Press, 1989), my major source for this section.

70. J.C. Beaglehole (ed.), *The Journals of Captain James Cook on his Voyages of Discovery: The Voyage of the Endeavour 1768–1771* (Cambridge: Hakluyt Society, 1968), pp. 44–5.

71. Beaglehole, *Endeavour Journals*, p. ccli.

72. Quoted in Smith, *European Vision*, p. 46.

73. Quoted on p. 105 of E.H. McCormick, *Omai: Pacific Envoy* (New Zealand: Auckland University Press, 1977), the best biography of Omai.

74. Quoted in O'Brian, *Banks*, pp. 183, 184.

75. Quoted in Carter, *Banks*, p. 131.

76. Quoted in O'Brian, *Banks*, p. 184.

77. Quoted in O'Brian, *Banks*, p. 186.

78. Quoted in Smith, *European Vision*, p. 82.

79. Quoted in McCormick, *Omai*, p. 227.

80. Quoted in Smith, *European Vision*, p. 116.

81. Quoted on p. 175 of Gascoigne, *Science in the Service of Empire*, a major source for this chapter.

82. Quoted on p. 109 of R. Drayton, *Nature's Government: Science, Imperial Britain, and the 'Improvement of the World'* (New Haven and London: Yale University Press, 2000), the other major source for this chapter.

83. Quoted in Gascoigne, *Science in the Service of Empire*, p. 32.

84. Quoted in Gascoigne, *Science in the Service of Empire*, p. 44.

85. Quoted in O'Brian, *Banks*, p. 224.

86. Quoted in Gascoigne, *Science in the Service of Empire*, p. 44.

87. Quoted in Gascoigne, *Science in the Service of Empire*, p. 46.

88. Quoted in R. Desmond, *Kew: The History of the Royal Botanic Gardens* (London: Harvill Press, 1995), p. 98.

89. Quoted in Desmond, *Kew*, p. 99.

90. Quoted in Desmond, *Kew*, p. 114.

91. Quoted in O'Brian, *Banks*, p. 235.

92. Quotations in Drayton, *Nature's Government*, p. 108, and Desmond, *Kew*, pp. 124–5.

93. Quoted in Drayton, *Nature's Government*, p. 118.

94. Quoted in Drayton, *Nature's Government*, p. 86.

95. Quoted in Gascoigne, *Science in the Service of Empire*, p. 140.

96. A.W. Crosby, *Ecological Imperialism: The Biological Expansion of Europe, 900–1900* (Cambridge: Cambridge University Press, 1986); Smith quoted on p. 195.

97. Quoted in Drayton, *Nature's Government*, pp. 87, 104.

98. Quoted in Gascoigne, *Science in the Service of Empire*, p. 180.

99. Both quotations from Drayton, *Nature's Government*, p. 105.

100. Quoted in Gascoigne, *Science in the Service of Empire*, p. 188.

101. Quoted in Desmond, *Kew*, p. 123.
102. Quoted in Gascoigne, *Science in the Service of Empire*, p. 186.
103. Quoted on p. 167 of Koerner, *Linnaeus*, a major source for this chapter.
104. Quoted on p. 253 of Gascoigne, 'Scientist as Patron and Patriotic Symbol', my other major source for this chapter.

ICONSCIENCE

THE ICON SCIENCE 25TH
ANNIVERSARY SERIES IS A
COLLECTION OF BOOKS ON
GROUNDBREAKING MOMENTS
IN SCIENCE HISTORY, PUBLISHED
THROUGHOUT 2017

The Comet Sweeper
9781785781667

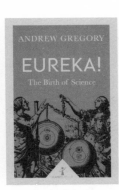

Eureka!
9781785781919

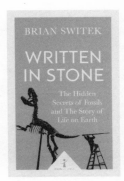

Written in Stone
9781785782015
(not available in North America)

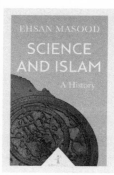

Science and Islam
9781785782022

Atom
9781785782053

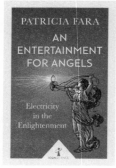

An Entertainment for Angels
9781785782077
(not available in North America)

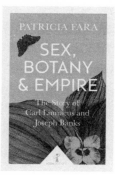

Sex, Botany and Empire
9781785782275
(not available in North America)

Knowledge is Power
9781785782367

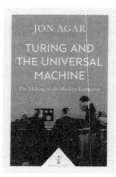

Turing and the Universal Machine
9781785782381

Frank Whittle and the
Invention of the Jet
9781785782411

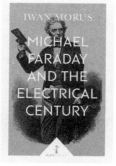

Michael Faraday and the
Electrical Century
9781785782671

Moving Heaven and Earth
9781785782695